CAMROSE LUTHERAN COLLEGE
LIBRARY

For Dr Johan Willgohs and the late Dr George Waterston who long nurtured the desire to see Sea Eagles return to Scotland; and to Harald Misund who made it possible.

JOHN A. LOVE

The return of the Sea Eagle

There be . . . things which are too wonderful for me . . .
The way of an eagle in the air . . .
Proverbs **30**: 18–19

CAMBRIDGE UNIVERSITY PRESS
Cambridge
London New York New Rochelle
Melbourne Sydney

Published by the Press Syndicate of the University of Cambridge
The Pitt Building, Trumpington Street, Cambridge CB2 1RP
32 East 57th Street, New York, NY 10022, USA
296 Beaconsfield Parade, Middle Park, Melbourne 3206, Australia

© Cambridge University Press 1983

First published 1983

Printed in Great Britain at the University Press, Cambridge

Library of Congress catalogue card number: 83-7325

British Library cataloguing in publication data
Love, John A.
 The return of the sea eagle.
 1. Sea eagle 2. Birds of prey – Scotland
 I. Title
 598'.91 QL696.A2

ISBN 0 521 25513 9

Contents

Preface ix
1 Introduction 1
2 Classification 5
3 Distribution 25
4 Breeding biology 47
5 Food habits 79
6 Persecution and decline 107
7 Conservation 131
8 Reintroduction 152
9 Release 176
10 Recolonisation 196
References 209
Index 219

Preface

For at least 4000 years men have trained hawks, falcons and eagles for hunting. Mongolian tribesmen may still employ Golden Eagles (*Aquila chrysaetos*) to hunt foxes and wolves for their pelts (Gordon, 1955), but by and large falconry has remained a leisure pursuit. Although practised in Britain by Saxon kings, it was not until the thirteenth century, during the Crusades, that it became fashionable. Traditionally, anyone was allowed to keep and train Kestrels (*Falco tinnunculus*) the yeomanry Goshawks (*Accipiter gentilis*) the aristocracy Peregrines (*Falco peregrinus*) but only kings and emperors held the honour of flying eagles. Golden Eagles were especially favoured, although it is said that King James I of Scotland (1406–37) flew a White-tailed Sea Eagle (*Haliaeetus albicilla*) when hunting Teal (*Anas crecca*) (Richmond, 1959).

We read how in Victorian times Sea Eagles might be kept as pets. One landowner on Skye had two which 'became his familiar companions, descending from a great elevation to join him on his walks, answering his whistle and retrieving game he had shot for their own larder' (Harvie-Brown & Macpherson 1904). About this time too, Dr Macgillivray and his two sons kept a Sea Eagle on the Isle of Barra, where it would 'follow them and hover round their heads, perform the most graceful aerial evolutions, and scream with delight as if it thoroughly understood and enjoyed the expedition; and when an unfortunate rabbit showed itself the eagle would swoop down upon it with amazing rapidity and power'. Ultimately the bird was mistakenly shot by a visitor from Glasgow 'to the great vexation of the entire household, with whom it had been a great favourite' (Gray, 1871).

Sea Eagles would appear to have made entertaining, if not always affable, pets. Gray recounted an amusing tale of one called Roneval, who would startle visitors to his master's house by swooping past the window. 'A well-dressed friend one day ventured to touch the bird with the point of his fashionable umbrella, which so offended Roneval's majesty that he flew at the offending instrument and literally smashed it, breaking the stick and tearing the silk to tatters – the owner, gladly escaping in unscathed broadcloth himself . . .' Roneval had a particular aversion to small boys yet would come to the call of the kitchen maid who fed him. Most of the time he did no more mischief than to occasionally help himself to a hen if his own dinner was not served up punctually!

'Eagles' flown to catch live fish were in all likelihood Ospreys (*Pandion haliaetus*), rather than Sea Eagles, although the eyries of both species were sometimes robbed to procure fresh fish for the table. One enterprising Highlander went so far as to put rings around the throats of eaglets so that he could be assured first of satisfying his own hunger (McIan, 1848). Osgood Mackenzie (1928) recounted how eaglets (in this case probably Golden Eagles) were tethered to delay their leaving the nest so that the adults continued to bring them food – hares, fawns, lambs and grouse which formed 'an agreeable variety at the shepherd's daily dinner of porridge and potatoes and milk'.

Sadly, however, man's preoccupation with eagles, whether Sea or Golden, was most often directed towards their destruction. The literature is overburdened with depressing detail of trapping, poisoning, shooting and egg-collecting – all of which will be described at length later in this book. The efforts of shepherds were inspired by their misguided belief that eagles were lifting live lambs, while wealthy collectors sought only to satiate their own hoarding instincts. One such was Edward Booth, devious and ruthless in his obsession, whose vast collection was ultimately bequeathed to Brighton Corporation. The aristocrat Charles St. John (1809–56) seems to have been little interested in eagles, preferring instead to pursue mercilessly the last of the Scottish Ospreys. Outstanding amongst such idle-rich collectors was John Wolley, who through his painstaking and detailed diaries contributed much to early ornithological knowledge. His accurate notes were later edited by the eminent ornithologist Professor Alfred Newton who published them as *Ootheca Wolleyana*. Mention should also be made of John Harvie-Brown (1844–1916) who travelled extensively in the Highlands and maintained prodigious correspondence with a vast network of colleagues and informants throughout the country (Love, 1982). His regional faunas remain invaluable reference works to this day. Notable amongst these informants was 97-year-old Duncan Macdonald, once a humble shepherd on the islands of Skye and Rum. Although Macdonald had destroyed many eagles in his long life, Harvie-Brown's notes reveal in the man a deep and passionate interest in the birds. Another contact for Harvie-Brown and Booth was the head stalker and forester of the Marquis of Breadalbane's Blackmount estate in Argyll, Peter Robertson.

Accounts by these early sportsmen, collectors and naturalists make entertaining reading but also provide an invaluable and irreplaceable source of information when researching into the past status in Britain and habits of birds such as the Sea Eagle. Finally the Sea Eagle was extirpated by the relentless endeavours of shepherds, keepers and collectors alike: to date it has failed to re-establish. Modern ornithological texts merely dismiss the bird as 'extinct' and rarely spare it further comment. Some even venture to condemn it out of hand as a lamb-killer, so as to justify our continued tolerance of the Golden Eagle. Since an attempt is now current to reintroduce the Sea Eagle to our shores, it is the purpose of this book to provide an up-to-date background of information upon the species. In this we are fortunate too in there being several major new studies undertaken abroad – but, sadly, only because there the species is becoming scarce under threats from persecution, habitat loss and toxic chemicals. One of the early pioneers of this research is Dr Johan Willgohs of Bergen University, who has made a lifetime study of the White-tailed Sea Eagle and whose monograph (1961) remains the standard work of reference.

After describing the distribution, breeding and food habits of the Sea Eagle, and tracing its demise in Britain, we shall consider conservation measures undertaken in Europe, before devoting the final chapters of the book to the reintroduction scheme being staged on the Isle of Rum in the Inner Hebrides. The lessons to be learnt from this innovative venture may prove relevant to the very survival of the White-tailed Sea Eagle on a global scale, and indeed of other threatened species. Inspiration for such a reintroduction attempt derived originally from Mr Pat Sandeman, the late George Waterston and Dr Johan Willgohs. This most recent project, the brainchild of Dr Ian Newton, is being undertaken by the Nature Conservancy Council (NCC) under the guidance of its Director (Scotland) Dr J. Morton Boyd and his Deputy Regional Officer Martin Ball. An Advisory Team was set up including Dr Boyd, M.E. Ball, P. Corkhill and R.T. Sutton (both ex-Chief Wardens of the Isle of Rum National Nature Reserve), Dr Derek Langslow (now the Team Chairman) and myself as Project Officer based on Rum – all six of us employees of the Nature Conservancy Council – together with Roy Dennis representing the Royal Society for the Protection of Birds (RSPB), Dr Ian Newton of the Institute of Terrestrial Ecology (ITE) and Dr Jeff Watson representing the Scottish Wildlife Trust (SWT). Latterly there has been financial input from the World Wildlife Fund (WWF) in both Britain and Norway, the RSPB and SWT. The project continues because of the tireless efforts of Captain Harald Misund in Bodø, Norway, the appropriate Norwegian authorities, Dr Willgohs and last but not least 120 Squadron of the Royal Air Force, Kinloss. Many other people have been involved at various stages since the project began in 1975, amongst them J. Bacon, R. Broad, C. Brown, Dr and Mrs J.L. Campbell, M.A. Carmichael, R. Coomber, C. Crooke, A. Currie, C. Duck, P. Duncan, Dr and Mrs A. Fowler, the Hon. Fiona Guinness, T. Jenkins, J.L. Johnston, M. Jones, Dr R Kenward, A. Leitch, I. MacKechen, I. Mackinnon and family, A. Maclennan, L. Macrae, J. Moore, J. Murray, D. Paynter, B. Philp, A. Race, Mrs Carol Scott, Mr and Mrs I. Simpson, G. Sturton, R.L. Swann, S. Taylor, C. Thom, B. Watt, G. Watt, the Hon. D.N. Weir, and M.J. Williams.

In the preparation of this book I have been privileged to receive help from many friends and colleagues, especially Dr H. Blair; Mrs Charlotte Buxton (Anglo-Saxon); Dr L.H. Campbell (for access to his late father's MS); R.H. Dennis; C. Fentzloff (Germany and captive breeding); Dr I. Fraser (place-names); Dr C. Harrison (fossils); Dr B. Helander (Sweden); Dr A. Ingolfsson (Iceland); Dr E. Jautsamo (Finland); Dr S. Karlsen (Norway); Dr J. Koivusaari (Finland); I. Lyster (Royal Scottish Museum); J. McCrum (Russian); I. MacIver (Latin); Dr F. Macleod (Gaelic); Dr M. MacDonald (German); D. MacQuarrie (Gaelic); Prof. H. Mendelssohn (Israel); W.R. Mitchell; D. Nethersole-Thompson; Dr I. Newton; T. Neumann (Germany); Dr G. Oehme (East Germany); Dr A. Pedersen (Icelandic and Iceland); Dr I. Pennie; R. Roxburgh; Dr A. Ruger (Germany); Dr Maeve Rusk; T. Soper; C. Spencer (Welsh); T. Stjernberg (Finland); M. Walters (British Museum); Dr G. Waterston; Dr A. Watson (placenames); F. Wille (Greenland and Denmark); Dr J.F. Willgohs (Norway) and Dr Y. Yom-Tov (Israel). Much pertinent research current in North America on Bald Eagles was brought to my attention by P. Nye and S. Postupalsky, amongst others. I am indebted to the libraries and librarians of the Nature Conservancy Council (Inverness, Edinburgh and Huntingdon) and the Scottish Ornithologists' Club (Edinburgh); and to Dr W.G. Hale, the Jourdain Society, and many museums – Arbroath (N. Atkinson), Bradford (Ms Margaret M. Hartley), Cambridge (R.D. Norman), Carlisle (D. Clarke), Delaware, USA (D.N. Niles), Dundee (A.B. Ritchie), Liverpool (J.G. Greenwood, P. Morgan), Manchester (Dr M.V. Hounsome), Perth (D. Bell), Sheffield (D. Whiteley), San Bernardino, USA (Ms Sharon Goldwasser) and Western Foundation of Vertebrate Zoology (L.F. Kiff). M.E. Ball, D. Buckland, Dr N.E. Buxton, R.V. Collier, R.H. Dennis, Dr D. Langslow, M. Marquiss and I.R. Taylor all read part or all of the manuscript; I greatly benefitted from discussion with Mick Marquiss. Mrs B. Sutton and Mrs M. Collier typed considerable sections of early manuscripts. Dr B. Helander, Prof. H. Mendelssohn, Dr. F. Wille and

Preface

Dr J.F. Willgohs kindly made available some of their excellent photographs, while the Hon. D.N. Weir lent me an early print of H.B. Macpherson's. Inevitably I must have omitted some people involved but to them apologies and together with those I have mentioned I extend very grateful thanks. Finally a special thanks to my wife Brenda for constant encouragement, comment and support.

1 Introduction

There heard I naught but seething sea,
Ice-cold wave, awhile a song of swan.
There came to charm me gannets' pother
And whimbrels' trills for the laughter of men,
Kittiwakes singing instead of mead.
There storms beat upon the rocky cliffs;
There the tern with icy-feathers answered them:
Full oft the dewy-feathered eagle screamed around.

The seafarer. (7th century)

The word 'erne' may be familiar to crossword addicts as an alternative name for the Sea Eagle. It derives from an Anglo-Saxon word meaning 'the soarer', and survives yet in some old placenames – Erne's Brae, Erne's Hamar, Ernesheugh and Earn Stack. (In modern Nordic languages an eagle is 'ørn', the Sea Eagle being 'havørn'.) Sadly the word has fallen from common usage but perhaps merits revival in view of the confusion which seems to persist. British ornithologists have settled for 'White-tailed Eagle', whilst their American counterparts prefer 'Gray Sea Eagle': possibly 'Erne' is too archaic, and 'Sea Eagle' too mundane. Some old accounts even called it 'Fish-eagle' and caused all too frequent confusion with the 'Fish-hawk' or 'Osprey'. In Welsh the species may be *Mor-eryr* (the big eagle) or *Eryr cynffonwyn* (the white-tailed eagle). Scots Gaelic aspires to several names – *Iolaire bhan* or *fhionn* (the pale or white eagle), *Iolaire ghlas* (the grey eagle), *Iolaire chladaich* (the shore eagle) or *Iolaire mhara* (the sea eagle); the bird also sports an attractive poetic name *Iolaire suile-na-grein* (the eagle with the sunlit eye), ultimate status as 'the true bird' being accorded to the Golden Eagle. The distinctive speckled or brindled plumage of the young White-tailed Eagle entitled it to be either *Iolaire bhreac* or *riabhach*.

No less an authority than Thomas Pennant, the eighteenth-century naturalist, classified the immature as the 'Sea Eagle' while the adult was considered a separate species which he termed the 'Erne' or 'Cinereous Eagle'. Even over its Latin name taxonomists have demurred; in 1758 Linnaeus had included the species in the genus *Falco* and it was not until 1809 that Savigny was to create the modern genus *Haliaeetus* for the sea/fish eagles, the White-tailed Sea Eagle itself being, not surprisingly, *Haliaeetus albicilla*.

Amid such confusion, which English name to choose? Personally I prefer the full title 'White-tailed Sea Eagle', to bring it into line with most others in its genus, as we shall see. If too cumbersome, the name can be conveniently shortened to 'Sea Eagle'. Throughout this book I adopt 'sea eagles' with lower case initials to refer collectively to the various species in the genus *Haliaeetus*. In everyday speech, however, one must be careful of one's enunciation, for more than once have I been accosted with the indignant rejoinder 'Why on earth do you want to reintroduce *seagulls?*'!

Another word of explanation – throughout the book I adhere to the ancient (Gaelic or Norse) spelling of Rum; not 'Rhum' the modern, but bogus, preference.

Since the dawn of history eagles have held special significance in the emotions and mythology of mankind. They represent the epitome of strength, keen sight and majesty. The eagle was adopted as the standard of Caesar and his legions, a practice since followed by many nations throughout the world. The magnificent Bald Eagle (*Haliaeetus leucocephalus*) was chosen by the United States of America as its national bird. Curiously in 1784 Benjamin Franklin lamented its being 'a bird of bad moral character; he does not get his living honestly; he is generally poor and very lousy'. Unjust comment indeed, especially since its only rival for this heraldic distinction had been the Golden Turkey! 'Such a vain and pompous fowl would have made a worse choice', countered the ornithologist Arthur Cleveland Bent in 1937. 'Eagles have always been looked upon as emblems of power and valour, so our national bird . . .', he added weakly, '. . . may still be admired by those who are not familiar with its habits.'

Besides being symbolic of power the eagle was held to be of religious significance. Its undisputed mastery in the air, for example, earned it a divine association with Jove or Jupiter, the classical god of the sky, or Taranis, his Celtic equivalent. In Nordic mythology, an almighty eagle, with a falcon perched on its forehead, sat on the topmost bough of an ash tree which itself signified the world of the gods (Davidson, 1964, 1969). The wings of this mythical eagle generated the four winds in the world of men. On a related theme the peoples of northeast Siberia reckoned that to kill an eagle would wrought fierce tempests from the skies (Armstrong, 1958). In Wales it was claimed that the great eagles of Snowdon bred gales and storms; indeed, an ancient name for this mystical mountain was *Creig ian 'r Eryri* (the rock of the eagles) (Gordon, 1955). Here the legendary King Arthur lies buried (one of the many resting places claimed throughout Britain!), his grave, according to Geoffrey of Monmouth, guarded by a pair of chained eagles. This same author described how 60 eagles would assemble annually at Loch Lomond in Scotland, to prophesy forthcoming events. In the twelfth century Geraldus Cambrensis avered how the eagle of Snowdon was possessed of oracular powers and foreshadowed war. She would perch on 'the fatal stone', sharpening her beak, before satiating her hunger on the bodies of the slain. Such a habit was known to the Anglo-Saxons who noted how, prior to the Battle of Maldon in 991 AD for example, 'a cry went up. The ravens wheeled above, the fateful eagle keen for his carrion'.

White-tailed Sea Eagles in particular had their own place in local tradition and medicine. The Anglo-Saxons believed that Sea Eagle bone marrow possessed miraculous curative properties (Armstrong, 1958) – it may well do so except that marrow is not to be found in the bones of any bird, let alone those of the Sea Eagle! The Faroese claimed that the claws were a cure for jaundice (Bijleveld, 1974) while to the Tartars the claws inflicted incurable wounds (not unreasonable due to the infection carried by a carrion-feeder such as the Erne).

Thus, over thousands of years man has erected around the eagle an exorbitant veil of mystique and legend, while even the phlegmatic ornithologist retains for it a charisma unrivalled except perhaps by the impressive Peregrine Falcon.

With some practice it is not difficult to distinguish the Golden Eagle from the White-tailed Sea Eagle. When flying, the Golden Eagle appears smaller, rarely exceeding 2 m in wing span, and is altogether more dashing and cavalier in its actions. While soaring, it holds its shapely, Buzzard-like wings upwards in a shallow 'V'. In silhouette, the neck is short, the tail long and square (fan-shaped when spread). At rest, the Golden Eagle possesses a neat, compact plumage, dark brown in colour, except for the golden plumes on head and neck; the tarsi are feathered right down to the toes. The juvenile appears almost black, hence its Gaelic name *Iolaire dhubh*, but has a conspicuous white base to the tail and white 'roundels' above and below each wing. These patches of white are gradually lost over several moults as the bird attains maturity.

In contrast, the Sea Eagle does not assume its characteristic white tail until adulthood, reckoned to be after at least 5 years. Immatures are dark brown, variously speckled with white, and have a dark tail, while the adult is greyish-brown, often fading to almost white on the head and neck. A yellow beak and eye and the brilliant white, crescent-shaped tail lend the bird a certain flamboyance. The tarsi are bare of feathers while the loose-feathered, almost untidy plumage convey a distinct vulturine aspect. The wings are massive, reminiscent too of a vulture and, as some would have it, like 'a flying door'! Its movements are distinctly ponderous with the wings held out horizontally flat when soaring, the tips often drooping slightly. This difference in the angle of the wings is not always an infallible distinction, and one must take into account size, shape, movement, tail, even habits; Sea Eagles, for instance, perch more readily on low and often flat ground, and seem to be more confiding in their attitude to man.

The White-tailed Sea Eagle is more vocal than the Golden Eagle, having a varied repertoire of sounds which are described in detail by Willgohs (1961), Fischer (1970) and Cramp & Simmons (1980). In brief the most common and versatile call is a rapid series of 15 to 30 short, shrill yaps which may be increased both in tempo and pitch. This is used as a greeting, in advertisement or in display, becoming almost a scream in aerial courtship. During copulation it has been rendered as a loud 'kee-kee-kee' fading to an almost inaudible 'ko-ko-ko'. While begging for food from the male, the female utters a long drawn-out whimpering, 'viieee-iiieee-iiieee', reminiscent of the persistent scream of an eaglet demanding to be fed. A bored or hungry youngster repeats a monotonous 'veee-veee' noise which becomes more extended and intense as soon as the parent returns. To an intruder an eaglet will hiss defiantly with open beak, sometimes beating one wing to inflict a painful knock with the bony carpal joint. When panic-stricken the bird will fall over on its back with half-open wings, to lash out with one or both feet. Mild alarm in both adults and young raises a hoarse 'kak-kak-kak', increasing in pitch and intensity according to the degree of threat. Rarely do adult Sea Eagles launch an aerial attack at an intruder in the nest.

Taxonomically, the Sea Eagle and Golden Eagle are not close but in order fully to comprehend their relationship and the similarities between the various sea eagle species we shall now consider their scientific classification, and indeed that of birds of prey in general.

2 Classification

> Methinks I see her as an eagle
> mewing her mighty youth
> and kindling her undazzled eyes
> at the full mid-day beam.
>
> John Milton (1608–74)

Eagles, hawks and falcons are normally classified within a single order – the Falconiformes. This view is now being challenged by recent biochemical research however. It may be that birds of prey are polyphyletic, having evolved from more than one ancestral stock (Sibley & Ahlquist 1972). Thus the apparent similarities between two of the three sub-orders – the falcons and the Accipiters – may be superficial, a result of parallel evolution towards a common predatory way of life.

The familiar term 'eagles' does not itself form a distinct taxonomic group, but rather embraces a spectrum of species not all closely related. Following the evolutionary sequence proposed by Amadon (Brown & Amadon, 1968), two-thirds of bird of prey species are contained within the sub-order Accipitres (Fig. 1). Of its three families, the Sagittariidae and Pandionidae contain but a single species each – the Secretary Bird (*Sagittarius serpentarius*), and the Osprey (*Pandion haliaetus*), respectively. The latter, because of its specialised fish-eating habits shares some adaptations with the sea eagles. Within the third family of Accipiters, the Accipitridae, two lines seem to have emanated from the primitive kites. One includes the snake eagles, harriers, hawks, buzzards and the true or 'booted' eagles (so

ORDER: FALCONIFORMES

Sub-orders: CATHARTAE (New World Vultures and Condors)
FALCONES (Falcons)
ACCIPITRES: families: Sagittariidae (Secretary Bird)
Pandionidae (Osprey)
Accipitridae sub-families:
   Milvinae (kites – 12 spp.)
   Aegypinae (Old World vultures – 14 spp.)
   Circaetinae (snake eagles – 12 spp.)
   Haliaeetinae:
     genera: *Gypohierax* (1 sp.)
     Ichthyophaga (2 spp.)
     Haliaeetus (8 spp.)
   Circinae (harriers, etc. – 13 spp.)
   Accipitrinae (hawks, etc. – 54 spp.)
   Buteoninae (buzzards, etc. –54 spp.)
   Aquilinae (eagles – 30 spp.)

Fig. 1. Classification of the Falconiformes

called because their legs are feathered down to the toes; this latter group contains the familiar Golden Eagle (*Aquila chrysaetos*).

The habits of some kites, and the Brahminy Kite (*Haliastur indus*) in particular, indicate a second line within the Accipitres which leads to the sea/fish eagles. This kite feeds near water where it will take crabs, small snakes and fish, as well as some carrion. Its courtship and plumage are reminiscent of the sea eagles (Fig. 2). Within this group are also classified the Old World Vultures, their affinities being suggested by the curious Vulturine Fish Eagle (*Gypohierax angolensis*) (Fig. 2). This bird is sometimes called the Palm-nut Vulture, reflecting the taxonomists' dilemma in deciding to which sub-family the bird belongs: if only for convenience it is most often classified with the sea eagles. Presumably as an adaptation to its feeding upon messy oil palm fruits, the bird's bright orange face is devoid of feathers – thus it bears a resemblance to the Egyptian Vulture (*Neophron percnopterus*) (Fig. 2).

Two species of fishing eagles from Southeast Asia are also classified with the sea eagles (Fig. 2). Where their ranges overlap, the Lesser Fishing Eagle (*Ichthyophaga nana*) frequents faster streams in forested areas and often at greater altitude; the Grey-headed Fishing Eagle (*I. ichthyaetus*) has a conspicuous white base to the tail and is the more vocal of the two species. Both nest near the tops of the trees and lay two to four eggs. Harrison &

Classification

Fig. 2. Species related to the genus *Haliaeetus*: Egyptian Vulture (Aegypinae), Brahminy Kite (Milvinae), Vulturine Fish Eagle (Haliaeetinae) and Fishing Eagles (Haliaeetinae).

Walker (1973) have described a third species, *I. australis*, from fossil deposits on the Chatham Islands in the Pacific.

A fossil, *Haliaeetus piscator*, has been described from Upper Miocene deposits in France (Brodkorb, 1964); all the remaining eight species of the genus *Haliaeetus* are still extant. They are, listed in roughly increasing order of size:

    Sanford's Sea Eagle, *H. sanfordi* (Fig. 4)
    White-bellied Sea Eagle, *H. leucogaster* (Fig. 4)
    Madagascar Fish Eagle, *H. vociferoides* (Fig. 3)
    African Fish Eagle, *H. vocifer* (Fig. 3)
    Pallas' Sea Eagle, *H. leucoryphus* (Fig. 5)
    American Bald Eagle, *H. leucocephalus* (Fig. 6)
    White-tailed Sea Eagle, *H. albicilla* (Fig. 7)
    Steller's Sea Eagle, *H. pelagicus* (Fig. 5)

Typically, the eight sea eagle species (Figs. 3–7) are aquatic or coastal in habits, and nest in trees or on cliff ledges or sometimes even on the ground. Usually two or three large white eggs are laid. Fish features to a greater or lesser extent in the diet but sea eagles may avoid possible competition with the Osprey by being less specialised and more ready to take waterbirds, carrion and sometimes mammal prey. They may also pirate food from other species, not least the Osprey itself.

Again typically, the *Haliaeetus* species display prominent patches of white in the plumage, notably on the tail but also on the head, neck or underparts (Fig. 8). In the adult the beak, cere and eye are usually yellow. The exception within the genus is Sanford's Sea Eagle which retains into adulthood an immature-type plumage of rufous and dark brown; its beak is black and its eye dark brown. The species seems to have derived from the White-bellied Sea Eagle, but having been isolated on the Solomon Islands, it occupies the niche of the true eagles – inhabiting dense, coastal lowland forest where it feeds on birds, such as pigeons, and mammals, especially phalangers. The nest and eggs of this rare and localised sea eagle have yet to be described.

Since the diet and breeding habits of the other seven species are so similar it is not surprising that in Europe, North America, Africa and throughout southeast Asia and Australia only one representative of the genus is to be found. Curiously, the coasts and vast river systems in South America are devoid of any sea/fish eagle (although there are fishing buzzards, *Busarellus nigricollis*). Nor has the genus *Haliaeetus* colonised Antarctica, New Zealand or the Pacific Islands (Fig. 9). Three species are to be found in the Palaearctic region, which might suggest this to be the ancestral home of the genus. Of these three the Steller's Sea Eagle is by far the largest – indeed the third largest eagle in the world (Brown 1976*b*): it is largely coastal in habits.

Classification

Fig. 3. African (left) and Madagascar Fish Eagles, *Haliaeetus vocifer* and *H. vociferoides*.

Fig. 4. White-bellied (right) and Sanford's Sea Eagles, *Haliaeetus leucogaster* and *H. sanfordi*.

Fig. 5. Steller's (left) and Pallas' Sea Eagles, *Haliaeetus pelagicus* and *H. leucoryphus*.

Fig. 6. American Bald Eagle, *Haliaeetus leucocephalus*.

Fig. 7. White-tailed Sea Eagle, *Haliaeetus albicilla*.

The smaller Pallas' Sea Eagle is less dependent upon water and occurs at altitudes of up to 5000 m in some parts of its range. A few White-tailed Sea Eagles have been known to have crossed the Bering Straits in recent times (Bent, 1961), presumably retracing the route by which this species once colonised North America where it became what is now the Bald Eagle. The habits and morphology of these two species are so similar that they are considered a superspecies. The Bald Eagle tends to be slightly shorter and narrower in the wing and has a less pronounced, wedge-shaped tail. Alaskan Bald Eagles are larger than their southern counterparts. As yet, the southern Bald Eagles have shown no inclination to cross the isthmus of Panama to colonise South America but instead move north to overwinter in the Upper States and Canada: as they return south in the autumn to take up breeding territories they are accompanied by the northern race seeking winter quarters! (Broley 1947).

The close affinities of the Bald Eagle and the White-tailed Sea Eagle are apparent not only in their almost identical habits but also in appearance. Whilst the body plumage of the latter is normally described as brown, its

Fig. 8. Plumage patterns on the undersides of *Haliaeetus* species (not to scale).

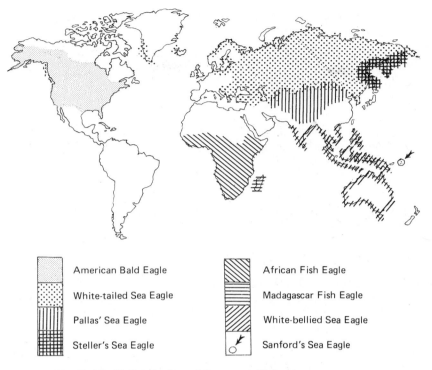

Fig. 9. World distribution of the genus *Haliaeetus*.

head and neck can appear conspicuously pale, almost white at times. Some individuals can be surprisingly pale all over – 'a fine silvery white, without the slightest admixture of brown' as Charles St. John (1849) described one eagle from Sutherland. He was told of another similar bird in its company, and it may have been one of these which came to be preserved in the museum of Dunrobin Castle. A Sea Eagle 'yellowish grey all over' found its way into a collection at Lews Castle, at Stornoway in the Outer Hebrides (Gray, 1871). Other like individuals from Sutherland were recorded by Wolley (1902) and Harvie-Brown & Buckley (1887). A silver female was shot on the Isle of Eigg in the Inner Herbrides in 1886 (Harvie–Brown & Macpherson 1904) while Harvie-Brown (1906) also recorded a pair of 'albinos' nesting in a tree in Loch Laidon, Perthshire, both parents and their young being 'alike of a pale dove colour or ash colour'.

This last record suggests there may be a genetic factor involved, although the common contention was that such pale individuals were of great age. Certainly the last Sea Eagle in Shetland, according to Lodge (1946) was 'quite white and looks as white as a gull when flying' (see Fig. 19): it had apparently been resident in the area for about 30 years. Prolonged exposure to the sun may bleach the feathers. Some Sea Eagles held in prolonged captivity during unusually sunny summers on the Isle of Rum, where they often shunned the available shade, assumed a pale plumage. After release this usually reverted to a more typical dark plumage at the next moult. Food seems not to be implicated at all, since these captives were provided with as natural a diet as possible.

When freshly moulted, the upper parts of a Sea Eagle's plumage often display a distinct violet gloss. This is thought to derive from the disintegration of small down feathers growing all over the body which, together with oil from the preen gland, may serve to enhance the waterproofing qualities of the plumage (Brown & Amadon 1968). The underparts lack this purplish cast and tend to be paler than the mantle. The blackish-brown primaries and secondaries on each wing are often washed with grey, and towards the base of their outer webs are flecked with buff. The tail coverts are white, variously mottled with brown and with diamond-shaped patches at the base and tips. These coverts conceal the brown bases of the otherwise white tail feathers.

The juvenile White-tailed Sea Eagle sports a coloration distinct from that of the adult. Its eyes are dark hazel brown while its beak is black; both gradually become yellow over 3 years or so as the bird matures. In its first year the bird is dark chocolate brown, especially on the head and neck, but more rufous on the mantle, rump, scapulars and wing coverts. More white is apparent on the underside, giving the juvenile a conspicuously speckled appearance. The white bases to any of the feathers flash prominently in a strong breeze, especially those at the back of the head. In flight the rufous-brown coverts on the leading edge of each wing contrast with the dark blackish-brown primaries and secondaries.

The plumage of any bird makes a considerable part of its bulk, weighing twice as much as the skeleton. Brodkorb (1955) diligently counted on a dead Bald Eagle 7182 vaned feathers, which collectively weighed 586 g. With an additional 91 g of down feathers, the plumage amounted to 17% of the bird's weight.

Feathers have to be renewed at intervals and it is not surprising that a bird the size of an eagle is unable to achieve this within a single year: moult ceases for 4 or 5 months in the winter, less if food conditions are particularly favourable. There has been no detailed study of the body moult of the Sea Eagle, but it is likely to be similar to that of the Golden Eagle. Jollie (1947) noted that his captive Golden Eagle showed the first signs of moult on March 11th, beginning with the head and neck which within a month looked 'positively unsightly'. The angle of the wings, the underside and all the middle coverts of the wings, together with the back and scapulars, were being renewed by early June. When feather loss ceased in early September, the belly and upper legs were only partially replaced. The next moult began on March 29th, starting on the head but spreading quickly to other parts of the body. This time the underwing coverts, belly, undertail coverts and tibial plumes were replaced for the first time, while parts of the head plumage, neck, back and scapulars were moulted again. Jollie considered that both moults had renewed almost the entire body plumage of his eagle, although he collected only 1850 and 2600 feathers (some of which had been moulted twice over the two seasons). If, in fact, like the Bald Eagle, the Golden Eagle possesses 7000 vaned feathers, only half the plumage may have been renewed.

White-tailed Sea Eagles possess 10 primaries (the eleventh being vestigial) and 17 secondaries on each wing. There is much individual variation in the sequence of moulting these. Cramp & Simmons (1980) outlined in brief a detailed study by Dr C. Edelstam of the Swedish Museum of Natural History, but a simplified moult sequence for a 'typical' Sea Eagle is presented by Forsman (1981). Forsman examined 65 skins, mostly of immatures, preserved in Swedish and Finnish museums. He found the first moult (when the bird is just over a year old) to commence in May/June continuing until October/November. Primaries 1, 2 then 3 (sometimes also 4) – being numbered from the body towards the wingtip – are shed. Secondaries are cast from 17 to 12 from the body out, while another centre of activity is initiated at the first secondary (see Fig. 10).

The second moult begins in March or April the following year, a couple of months earlier than the first. Primary moult continues where it had left off, with the renewal of the fourth, fifth and sixth. The second and third secondaries are replaced and numbers 11 and 10 continuing the previous year's moult; a new centre is begun with the loss of the fifth and sixth while the innermost three (17 to 15) are replaced for a second time.

On the third moult, primaries 7 and 8 (sometimes also 9) are renewed, and the process begins anew at the first primary. The outermost tenth (and perhaps also the ninth) are still the original first year feathers and are therefore exceedingly worn and faded. The secondary moult continues from its three centres of activity, with the loss of the fourth, the seventh, eighth and ninth, and numbers 14 and 13. By now all the secondaries will have been renewed at least once.

Thus the moult of the primaries has proceeded over several years, in waves from the carpal joint outwards, in a process termed 'serially descendant' (Cramp & Simmons, 1980). The outermost secondaries are also moulted in a serially descendant manner, but the innermost are lost 'ascendantly', from the body out towards the carpal joint of the wing. An

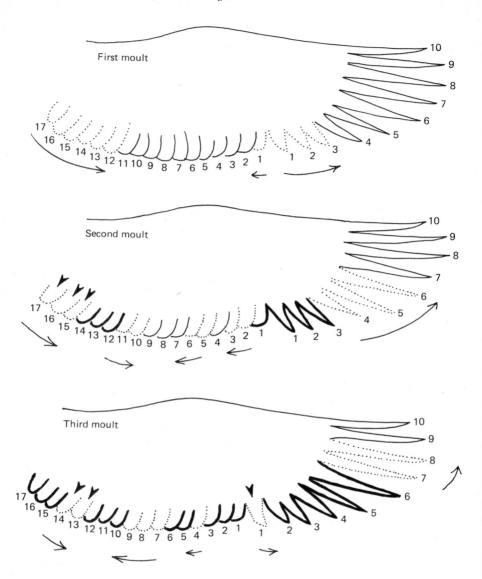

Fig. 10. Simplified primary (1–10) and secondary (1–17) moult in White-tailed Sea Eagles (after Forsman, 1981). Dotted feathers are freshly cast, heavily outlined feathers were renewed the previous moult. Long arrows indicate the direction of moulting. Feathers marked with arrow heads have been renewed twice.

'average' wild Sea Eagle, according to Forsman, may lose three or four primaries and six or seven secondaries at each annual moult. Captive eagles, however, tend to be kept under optimum conditions and in any one season seem to replace more feathers than wild birds. Three captive Sea

Eagles on Rum lost about five or six primaries from each wing during their second moult – I was unable to study satisfactorily their secondary moult. By collecting the shed feathers each day I was able to ascertain that primaries tended to be lost as pairs – the equivalent feathers from each wing at intervals of 2–6 days (but sometimes as much as 25–60 days). A few primaries were lost from one wing only, their equivalents on the other wing being retained until the following year.

It is obvious that second-generation secondaries tend to be several centimetres shorter (and less pointed) than the first year's. Thus, in 2 and 3 year olds those feathers they retain from the first year protrude slightly beyond the trailing edge of the wing. This can be a useful aid in ageing Sea Eagles in the field.

There are 12 tail feathers (retrices), numbered as pairs from the centre of the tail outwards. My captive Sea Eagles provided some data on their moult but it seems to have been even less predictable than the wing moult. In general the innermost (first) pair tended to be shed first, followed by the outermost (sixth) or sometimes fifth pair. The other pairs tended to be shed more or less at random. As with primaries, both members of each pair were often dropped within a few days of one another, but sometimes as much as 10 or more weeks apart. This followed roughly the same pattern as Jollie's captive Golden Eagle, which, however, replaced four pairs in its first moult and seven feathers in the second – the three innermost being replaced twice. Observations on a single captive juvenile Sea Eagle on Rum indicated that nearly all of the tail was renewed in the first moult, while three older captives moulted only seven or eight feathers in their second moult. Feathers retained from the previous year, together with any damaged ones, were often amongst the earliest to be cast. The renewal of any remiges which dropped out at the root during the year began almost immediately but, curiously, they assumed the characteristics (length, shape and coloration) of the next year's. It required about 11 weeks to completely regrow a tail feather, and 'growth marks' visible on the underside of the rachis indicated a daily rate of about 7 or 8 mm. Again, Jollie found that while wild specimens might replace as few as two retrices, captive birds might replace eight or nine. This stresses the need for caution in extrapolating data from captive birds to a wild situation.

The functional significance of replacing flight and tail feathers gradually, and of having an incomplete annual moult (despite several simultaneously active moult centres) is clear: a bird so dependent upon flight to secure its prey must retain an adequate degree of symmetry in wings and tail, and a sufficient area to provide adequate lift and manoeuvrability. One-year-old juveniles show more pointed secondaries, giving their wings a distinct serrated trailing edge; their tail is also longer and more rounded because the outermost pair (and to some extent the next outermost) grow in some 10% shorter after the first moult. These features may serve some aerodynamic function but, together with the dark immature plumage, provide an appearance and silhouette instantly recognisable from those of an adult Sea Eagle (Fig. 11).

The four captive Sea Eagles on Rum have allowed some useful observations to be made on the coloration of the tail feathers over three successive moults – a useful indicator of a bird's age (Fig. 12). The basal part of each retrice (normally one-seventh to one-third of its length) is brown, and in the first year this continues up the whole of the outer web. The inner web is mostly buff with dark flecks and tipped in dark brown. In successive moults this buff area becomes paler and more extensive. Simultaneously the outer web begins to lighten in colour towards the tip and also from the margin with the permanent brown base. The relative amounts of brown at the base and white on the rest of the retrice varies between individuals, but for each remains consistent throughout its life. Thus adult birds can be individually recognised, much as Opdam & Muskens (1976) have described in *Accipiter* hawks. Thomas Neumann in West Germany collects moulted tail feathers (Fig. 13) in each White-tailed Sea Eagle territory so that the fortunes of each individual in his study population can be followed from year to year: this precludes the need for any sophisticated marking technique.

Dark tips may be retained on the tail feathers into the fourth or even fifth year, but at no stage do they ever resemble the continuous, broad black terminal band of the juvenile Golden Eagle. Without any dark pigmented tips, the white tail of an adult Sea Eagle seems more vulnerable to abrasion so that, especially towards the end of winter, it can appear conspicuously short.

There are no obvious sexual differences in plumage amongst adult Sea Eagles, although my close association with many juveniles has indicated several minor features of some value. Males tend to be darker brown and less speckled on the head and neck than do females. The head and neck plumes of the male appear to be shorter so that when sleeked down (as when frightened) the slighter build of the skull is accentuated, giving the crown an almost angled appearance. Males tend to be more highly strung and, depending on their nervous state, possess a higher-pitched voice. Size is the most useful feature in distinguishing the sexes. The wings and tail of females are some 6 or 7% longer than those of males, the beak some 8–10% longer and the tarsi about 14% thicker: females are about 20% heavier than males (Table 1). Weight is, however, a highly variable statistic when one considers

Fig. 11. Juvenile White-tailed Sea Eagle showing characteristic silhouette of tail and trailing edge of wings.

that a kilogram or more of food can be consumed at a single meal: weight can also vary both daily and seasonally. Wing length is not always easy to measure accurately in such a large bird as a Sea Eagle and where the wing tip is subject to abrasion or damage. Beak length is dependent upon the rate of wear at the tip. Two measurements have been found to be most useful in distinguishing the sexes – tarsus thickness and beak depth (see Figs. 70 and 71). If these two statistics are plotted one against the other, a good separation between males and females can be obtained with almost no zone of overlap (Fig. 14). The sexes of the individuals involved were determined using subjective assessments of voice, behaviour, appearance and size; in four individuals I was able to confirm my judgement by autopsy. Helander (1981c) has similarly investigated measurements of Swedish Sea Eagles and has confirmed the usefulness of tarsus thickness; he found wing length to be useful in ageing nestling eagles.

Dimorphism in size is less pronounced in the Sea Eagle, being a species feeding on less active prey, than amongst some other birds of prey. True carrion-feeders exhibit almost no size dimorphism, while a species such as the Sparrowhawk (*Accipiter nisus*) specialising in fast-moving bird prey, can have females twice as large as the male (Newton, 1979). The size dimorphism demonstrated by predatory birds (and in this respect owls and even skuas can be included) is reversed, in that it is the female which is the larger. Newton (1979) discusses several ecological and behavioural explanations which have been postulated to account for this, but no single explanation seems satisfactory.

There have been no detailed biometric studies of the White-tailed Sea Eagle throughout its extensive range but it is possible that there may be a clinal increase in size from north to south (as indicated by egg dimensions in Table 5). Such a cline is present in the Bald Eagles of North America (Table 2). The largest Sea Eagles are known to occur amongst the isolated population in Greenland. The following chapter will describe in more detail the geographical range of the species.

20 The return of the Sea Eagle

Fig. 12. The female Sula, showing characteristic flight silhouettes at various stages in development. (a) Third year; typical broad-winged silhouette with wedge-shaped tail – little white obvious as yet, Rum, April 1979. (b) Fourth year; showing pale beak and eyes, Rum, March 1980. (c) Fourth year, premoult; showing abraded primaries and tail,

Classification

Rum, April 1980. (*d*) Fourth year, premoult; taken on same day as (*c*) but little white obvious on the tail when it is not widely fanned, Rum, April 1980. (*e*) Fourth year; in heavy moult, Rum, June 1980. (*f*) Fifth year; showing a few dark tips retained on the otherwise white tail; a few secondaries in moult, Rum, July 1980. (Photos: J.A. Love)

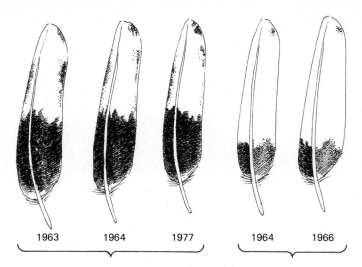

Fig. 13. Outermost tail feathers of two adult White-tailed Sea Eagles, showing the annual consistency of pattern but differences in pattern between individuals. (From the collection of T. Neumann, West Germany.)

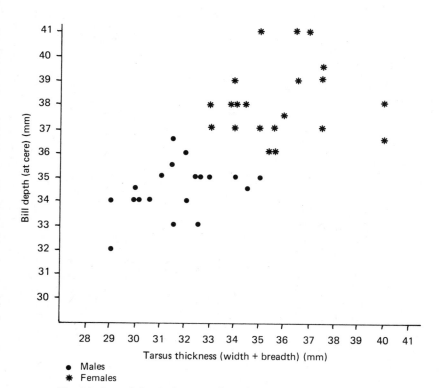

Fig. 14. Sexual dimorphism in the White-tailed Sea Eagle, using bill depth and tarsus measurements (in mm) as an aid in sex determination. Birds measured have been sexed on appearance, behaviour and voice; in four this was later verified *post mortem*.

Table 1. *Measurements and weights of White-tailed Sea Eagles*

Measurements	Age	Male	Female	% ♀ larger than ♂	Source
Wing length (mm)	Ad	606±26(19), 552–640	668±26(14), 621–715	10.2	Cramp & Simmons, 1980
	Imm	645±22(24), 610–695	685±27(29), 615–740	6.2	Cramp & Simmons, 1980
Wing span (mm)	All ages	2262(5), 2200–2336	2370(8), 2260–2470	4.8	Willgohs, 1961
Tail length (mm)	Ad	280±19(17), 254–331	305±19(15), 276–330	8.9	Cramp & Simmons, 1980
	Imm	322±24(23), 280–370	345±18(28), 315–380	7.1	Cramp & Simmons, 1980
Hind claw (mm)	Ad	39.4±1.6(13), 37–41	42.3±2.3(11), 39–46	7.4	Cramp & Simmons, 1980
Tarsus length (mm)	Ad	94.5±2.0(10), 92–97	96.5±2.2(7), 95–101	2.1	Cramp & Simmons, 1980
Tarsus depth (mm)	Imm	16.9±1.1(20), 15–19	19.0±1.3(24), 17–22	12.4	This study
Tarsus width (mm)	Imm	15.1±0.8(19), 13.5–16.5	17.2±1.2(23), 15.5–19.5	13.9	This study
Bill length (mm)	Ad	52.6±2.8(17), 47–58	58.6±3.8(12), 51.5–63.5	11.4	Cramp & Simmons, 1980
		54.0±3.0(3), 51–57	60.5±1.8(7), 58–63.5	12.0	Willgohs, 1961
	Imm	52.9±2.0(20), 50–56	56.7±3.0(22), 51–61	7.2	Cramp & Simmons, 1980
		53.8±1.6(11), 51–56	58.0±2.6(13), 52–61	7.8	Willgohs, 1961
		53.5±2.0(24), 49–57	58.0±1.6(30), 55–62	8.4	This study
Bill depth (mm)	Ad	32.3(1)	39.0±6.2(3), 32–44		Willgohs, 1961
	Imm	34.2±1.3(6), 32–36	38.0±2.6(10), 34.5–44	11.1	Willgohs, 1961
		34.8±1.2(24), 32–37	37.8±1.8(30), 34–41	8.6	This study
Weight (kg)	Ad	4.0±0.8(12), 3.1–5.4	5.6±1.0(18), 4.1–6.9	40.0	Cramp & Simmons, 1980
	Imm	4.2±0.7(10), 3.0–5.0	5.4±1.0(19), 3.2–7.5	28.6	Cramp & Simmons, 1980
		4.5±0.5(5), 3.6–5.0	5.6±0.6(9) 4.6–6.5	24.4	Willgohs, 1961
		5.0±0.6(24), 4.0–6.5	6.1±0.4(28), 5.3–6.8	22.0	This study

Males are smaller than females. Immatures (Imm) have longer wings and tails than do adults (Ad). Data in Cramp & Simmons (1980) are gathered from various sources including Willgohs (1961), but in places some of Willgohs' data are included to compare with equivalent data from Rum since they both derive from the same population in Norway. The captive birds on Rum are heaviest (on release), doubtless being of a healthier weight than museum specimens (at the time of collection). Data given are mean ± standard deviation (sample size) range.

Table 2. *Comparison of wing lengths and weights of Bald Eagles and White-tailed Sea Eagles*

Race	Male	Female	Source
WING LENGTH (mm)			
H. l. leucocephalus	529 (?), 515–545	577 (?), 548–588	Friedman, 1950
H. l. alascanus	589 (?), 570–612	640 (?), 605–685	Friedman, 1950
	619 (8), 577–650	655 (8), 620–705	Dementiev & Gladkov, 1951; Southern, 1964
H. a. albicilla	595(23), 575–625	653(17), 635–690	Dementiev & Gladkov, 1951
	641(26), 590–695	685(37), 650–740	Willgohs, 1961; Fischer, 1970
H. a. groenlandicus	647(30), 624–665	691(27), 660–727	Salomonsen, 1950
WEIGHT (kg)			
H. l. alascanus	4.1 (4), 3.9–4.5	5.6 (7), 4.8–6.6	Dementiev & Gladkov, 1951; Southern, 1964
H. a. albicilla	4.4(12), 2.8–5.4	5.6(22), 4.6–6.9	Willgohs, 1961; Fischer, 1970

Data are given as mean (sample size), range.

3 Distribution

> Stray into districts such as these, faraway from man's haunts and industries, and there it is that the Sea Eagle will come out of the mountain mists yelping fright at your intrusion, and sail proudly outwards displaying his grand powers of wingmanship to your astonished and delighted gaze.
>
> Seebohm (1883)

In compiling a distribution map one is frustrated by the paucity of information from many parts of the world. The White-tailed Sea Eagle breeds throughout much of the Palaearctic zone (Fig. 15) but only for the countries of Western Europe do detailed descriptions exist. Thus, while one can carefully outline the discontinuous and patchy distribution of this much reduced species in the west, one can only shade in a vast tract of colour across the USSR and the near east. A cursory glance at the completed map therefore conveys the erroneous impression that while scarce in Europe, the bird can be in little danger in view of its widespread distribution elsewhere. This state of affairs highlights the need to assimilate whatever knowledge one may possess on status, past and present.

The Greenland race, *H.a. groenlandicus*, for instance, was at the turn of the century to be found from Cape Farewell along the west coast to Disko Bay, about 69° North. It may have been twice as numerous as it is now – none now breeding north of Artersiorfik Fjord (67°30′ North). There seems to have been a similar contraction of its range to the south, so that the highest density is now to be found in the Frederikshab region (61–62° North). Hansen (1979) has estimated the present population to number some 85 to 100 pairs. Collecting has long been a problem and for many decades it has

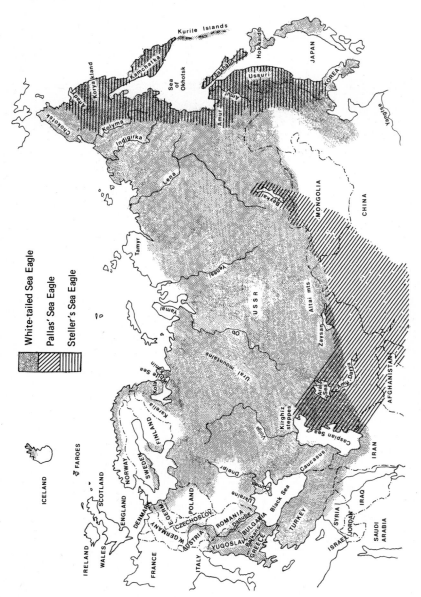

Fig. 15. Distribution of White-tailed Sea Eagle.

been illegal to export White-tailed Sea Eagles or their eggs from Greenland. The species only received full legal protection as recently as 1973, however. Now the main threat is from sheep farmers who accuse the eagles of killing their lambs. Danish ornithologists have embarked upon a detailed study of the eagle and tend to the view that the recent excessive losses of sheep and lambs is due to a deterioration in climate (Wille, 1977). They have launched a campaign to educate the public that this magnificent eagle is under dire threat but many are still shot or trapped by sheep farmers – at least one bird fell to the gun of a local policeman! Salomonsen (1967) estimated that over a third of the eagle population was being destroyed thus, but fortunately many nesting areas remain remote and their production of young remains encouragingly high.

It was doubtless from Iceland that the Greenland colonists derived, but there is now unlikely to be much contact between the populations of these two countries since during the early part of this century the Iceland Sea Eagles were reduced almost to the point of extinction from which they have never really recovered. Gudmundsson (1967) considered that some 100 or 200 pairs may once have nested. By 1870 this was reduced to about 70–80 pairs, the rate of decline thereafter accelerating until no more than 20 pairs were left by 1910. The full legal protection instigated 3 years later halted any further decline but has not encouraged much recovery. There may now be fewer than 40 individuals left and only some 10 pairs may now nest with any success, all in the northwest of the country. The eagle became a target for persecution by man because it was said to prey upon lambs and on Eider ducks whose down is collected on a commercial scale in Iceland. Arctic Foxes were considered another serious menace and until 1964 poison baits were laid for them which accounted also for many eagles (Ingolfsson, 1961; Gudmundsson, 1967; Cramp & Simmons, 1980).

A few Sea Eagles bred in the Faroes until the middle of the eighteenth century when they finally became extinct there. Names such as 'Årnafjord' and 'Årnafjall' are all that remain to indicate their former haunts, and few vagrants are ever seen – the last, according to Williamson (1970) in 1902.

In France a number of Sea Eagles winter regularly on Etang de Lindre in Lorraine (Bijleveld, 1974) and stray birds have at times been recorded in Spain (Bent, 1961), Portugal and Malta (Cramp & Simmons, 1980). A few pairs may have bred in Corsica until the 1940s; one was seen on that island in 1959 by a man rejoicing in the name of 'Ern'! (Bijleveld, 1974). The species would appear to have become extinct on the neighbouring island of Sardinia around 1960 (Cramp & Simmons, 1980).

There was an eyrie on Lake Menzaleh in Egypt during the nineteenth century (Meinertzhagen, 1930). Elsewhere in North Africa rare individuals have been reported from Algeria and Tunisia; they have also been reported from the Canaries, Cyprus and Lebanon. Two pairs frequenting the Jordan valley in Israel ceased to breed in the early 1950s due to the effects of agricultural chemicals (Yom-Tov, personal communication). The Sea Eagle is said to breed in central Iraq, although of this there has been no recent confirmation (Cramp & Simmons, 1980). The Sea Eagle also occurs along the shores of the Caspian Sea in Iran (Vaurie, 1965) and it formerly bred in northern Syria (Cramp & Simmons, 1980).

The wetland areas of western Turkey now support only 20 or 30 breeding pairs and they are currently threatened by encroachment upon their habitat by man (Acar, Beaman & Porter, 1977). In Greece the species was once very common but now only 20–30 pairs breed, mostly on the Evros delta near the Turkish border but also on the islands round Thessaly, eastern Macedonia and Thrace. Tragically, the once rich wetlands of Greece are now threatened with drainage and reclamation (Vagliano, 1977). Early this century White-tailed Sea Eagles bred in Albania but their present status is unknown (Cramp & Simmons, 1980). The species may now have disappeared from Bulgaria where it had formerly been common along the Black Sea coast; it is possible that one or two pairs may yet exist on the banks of the Danube (Bijleveld, 1974). Around 1960 some 40 pairs may still have bred on the Danube delta, although in Romania the species was described in the nineteenth century as 'unimaginably numerous', a severe reduction took place after 1924. The present population, both along the Black Sea coast and on the Danube delta, is no more than 10–15 pairs (Bijleveld, 1974), while Puşcarin & Filipaşcu (1977) consider the species to be on the verge of extinction in Romania. Yugoslavia also held considerable numbers in the past and some 80 pairs still bred in the early 1950s. The numbers have since been reduced by the use of strychnine (against wolves) and by the effects of toxic chemicals. Only 10–15 pairs now occur (Bijleveld, 1974) mainly in the Vojvodina. The total fluctuates from year to year as several pairs utilise alternative eyrie sites across the border in Hungary. The banks of the Hungarian Danube were once a stronghold but by the 1950s the numbers had declined to 10–15 pairs because of poisoning and the destruction of forests; at present the Sea Eagle seems to be on the verge of extinction in Hungary. Indeed none may have bred in 1973 but five eyries were located 2 years later (Bijleveld, 1974; Bécsy & Keve, 1977). More are to be found in Hungary in the winter, especially on the Hortobagy where up to 35 have been known to roost in one wood (Fintha, 1976; Bécsy & Keve, 1977). A few pairs also bred along the Czechoslavakian Danube and other rivers in the country, but this is now down to a single pair which is strictly protected (Sládek, 1977).

At one time small numbers were to be found in Austria but after 1859, when one of the last females was shot at the nest, breeding in that country became more sporadic. One pair attempted to breed each year until 1946 but, although some individuals have summered near Vienna and Orth in recent years, none has become re-established. There has also been a marked reduction in the wintering population (Bijleveld, 1974; Bauer, 1977; Cramp & Simmons, 1980).

In West Germany, Sea Eagles were to be found in Bavaria – until the middle of the nineteenth century – and in Schleswig-Holstein. A decline set in, until in 1920 it was claimed that the last pair bred in Schleswig-Holstein. Hitherto it has been assumed that recolonisation took place from Mecklenburg in East Germany in 1945 but a few pairs may have nested in the west prior to this (Ruger, 1981). They increased after the Second World War until, in the 1950s, some eight pairs bred annually. The next decade saw a fresh decline induced by the combined effects of disturbance, collection, persecution and toxic chemicals. Protection measures initiated since 1969 have stablised the population at four pairs – since 1976, a fifth pair has defected to the East, utilising an alternative eyrie just across the border. Their breeding output remains low although it has been augmented in recent years by captive breeding and adoption techniques (described at greater length in Chapter 7).

Last century the White-tailed Sea Eagle was exceedingly numerous in what is now East Germany, both in the lake areas inland and along the Baltic coast. Around 1913 the population was severely depleted by persecution. Protective measures and the abolition of bounties after the First World War initiated a recovery, with the numbers of breeding pairs achieving a peak by the 1950s. Illegal shooting and poisoning has continued but with little effect on the population compared with that induced by pesticides. Oehme (1961) and Bijleveld (1974) reckoned that some 30 pairs were still breeding in Brandenburg, around Berlin, mainly in the Potsdam and Frankfurt/Oder districts, while two pairs were to be found between Magdeburg and Dresden to the southwest. It is, however, in the Mecklenburg area that the bulk of East German Sea Eagles are to be found – some 80 to 90 pairs (Oehme, 1961; Bijleveld, 1974; Dornbusch, *per* R.H. Dennis, personal communication). Ruger (1981) details 43 pairs in the lake areas of Neu-Brandenburg, a further 22 pairs near Schwerin and another 23 pairs (possibly now declining) nearer the coast around Rostock.

Whereas at one time the species occurred throughout Poland, it disappeared early from the central parts and Silesia. Bogucki (1977) estimated 43–50 pairs, mainly on the Baltic coasts of western Pomerania, with some on the Masurian lakes. Isolated inland pairs are to be found in the provinces of Wroclaw, Lublin, Poznan and Zielona Gora. There are no current data available on breeding success (Bijleveld, 1974; Bogucki, 1977).

Nor is there much information from the Baltic states of Estonia, Latvia and Lithuania except that there might have been a slight recovery in numbers after the Second World War, possibly to about 20 occupied territories. Pesticides have now reduced breeding success so that there are only five pairs which may breed (Bijleveld, 1974). Galushin (1977) estimated that some 50 pairs of Sea Eagles may nest in the central part of the western USSR but Bijleveld would consider this too high a figure. A decline had set in by the early decades of this century and the species has now been placed on the White (rare) list of the USSR Red Data Book on endangered species. The distribution in the USSR has been outlined by Dementiev & Gladkov, (1951), Kozlova (in Bannerman, 1956) and by Vaurie (1965) on whose data the following account is based. The species breeds (probably in low numbers) around the Black Sea but may now have ceased to breed in the Crimea. It is found along the lower reaches of the Dniepr river in the Ukraine (although only two pairs were recorded there in 1937). It is common still around the Caspian Sea, on the Volga and on the Kirghiz steppes. Breeding is sporadic in Turkestan where the species extends westwards through the lower reaches of the Amu-Darya and Ili rivers to the Zaysan depression, the Altai mountains, northwest Mongolia, Transbaykalia, Amurland and northern Manchuria. According to Vaurie it once bred along the lower Yangtze river in China. Along this southern limit of the Sea Eagle's range it may be found alongside Pallas' Sea Eagle. On Lake Baykal, for instance, local ornithologists knew of at least 12 pairs of White-tailed and 6 pairs of Pallas' Sea Eagles (R.H. Dennis, personal communication). It would appear also to overlap with Steller's Sea Eagle along the Pacific coast. Unfortunately there is little information on these species where they are sympatric but it would seem that the White-tailed Sea Eagle is less migratory than the other two. However, at least some White-tailed Sea Eagles are known to move south for the winter concentrating off Japan, Korea, Ussuriland, Turkestan and by the Caspian Sea. It may also be that Steller's Sea Eagle is more confined to the coast and because of its greater size can feed on larger fish and birds.

On 15th February 1980, Mori & Nakagawa (1981) counted 383 Steller's and White-tailed Sea Eagles wintering on the Shiretoko Peninsula of Hokkaido, Japan. The area held about 15 breeding pairs of the latter in the late 1970s, but the numbers have since declined (N. Ohtaishi, personal communication).

The eastern limit of the White-tailed Sea Eagle's range extends from eastern Hokkaido in Japan and the shores of the Sea of Okhotsk from

Sakhalin, the Kurile Islands and the Kamchatka peninsula to Koryakland, Anadyrland and the Chulotski peninsula. From the northeast of Siberia the Sea Eagle extends from the tundra zone proper along large forested river valleys. It is reported to be abundant around the Kolyma, Indigirka and Lena rivers to about 72° North and possibly even to 75° North on the Tamyr peninsula. It is common on the Ob river extending to 70° North at the mouth of the Yenesei river and on the Gydan and Yamal peninsulas. Around the White Sea it is said to be the most abundant bird of prey, being found both on the coast and on the inland lakes of Kanin, Kola and Karelia. According to Flerov (1970) however, the breeding success in Kandalaksha Bay of the White Sea is low, often due to periods of cold weather in the spring and to human disturbance.

In Finland in the last century, the Sea Eagle was reduced by persecution to around 20 pairs by the 1920s. Subsequent to its being protected the population doubled but in the early 1970s the recovery was halted by environmental pollutants. Only a few pairs now attempt to breed and their reproductive output remains low. In 1980 there were about 45 territories occupied, not all of them by pairs however: 15–20 on the Åland Islands, nine on the rest of the Archipelago Sea, 15 on the Quarken Straits in the Gulf of Bothnia, and 6 to 10 pairs in Finnish Lappland (Henriksson, Kappanen & Helminen, 1966; Bergman, 1977; Stjernberg, 1981). The situation was similar in Sweden in that the species was persecuted last century so that by 1924, when it became protected by law, only 20–30 pairs survived. A slight increase then occurred until the pesticide era much diminished the reproductive output of the eagles. In 1964 about 82 pairs bred along the Baltic coast from Småland to Vesterbotten, although this has been reduced in recent years to 50–60 pairs. A further 20 pairs occur in inland areas of Swedish Lappland, little affected by toxic chemicals but perhaps subject to some human disturbance and persecution (Helander, 1975, 1981a).

As recently as 1880, as many as 50 pairs were apparently to be found in Denmark but by 1911 only a single pair remained in the northern part of Jutland; the sole surviver of this pair was shot in 1918. During the 1950s some nesting attempts were made again in southern Jutland, in Lolland and on Seeland but only the latter pair succeeded, in 1954. A pair also bred in 1980 but the eggs, although fertile, failed to hatch (F. Wille, personal communication). Up to 10 individuals may winter in Denmark (Bijleveld, 1974; Dyck, Eskildsen & Møller, 1977).

At the end of last century the White-tailed Sea Eagle bred along the entire coast of Norway from the Swedish border in the south to the Finnish (now Russian) border in the north. A noticeable decline then occurred, mostly in the south, so that at the present time breeding extends from the Sogn district northwards. A census by Willgohs (1961) counted some 350 pairs between 1956 and 1960. Full protection was given in 1968 which appeared to check any decline and even permitted local increases further north. Another census in 1974–76 (Norderhaug, 1977) revealed about 450 pairs, no doubt due to an improved coverage in the census but also, it is thought, reflecting a true increase in numbers. Over one-third of the Norwegian population is to be found in Nordland, with good numbers also in Troms and Finnmark where the broken and remote coastline still defies a complete coverage. Illegal shooting and collecting takes place on a small scale and minute traces of toxic chemicals are now being detected in dead eagles and eggs (Norderhaug, 1977), but on the whole the breeding success remains high.

Norway thus remains one of the last safe strongholds of the White-tailed Sea Eagle in Western Europe. It possibly holds one-half of the entire Western European population, which may amount to no more than 1000 pairs; but just how many remain throughout the USSR is impossible to assess. If the pattern exhibited in most other countries can be taken as representative, some decline must have occurred last century due possibly to loss of habitat and human persecution, and latterly even to pesticides. The enclosed seas such as the Baltic have been demonstrated to be the most vulnerable in this latter respect so that it is likely (and indeed there is some evidence) that the Caspian and Black Seas and Lake Baykal in the USSR are similarly affected by pollution. The major river systems of Siberia are an important habitat; it should be cause for much concern that the Soviets are contemplating drastic modifications in order to divert water for irrigation in arid areas further south.

In summary, habitat loss and persecution have seriously diminished the numbers of Sea Eagles in most European countries, with the hazards of indiscriminate pesticide use checking any recovery which might have been incurred from bird protection laws. It is perhaps surprising then that the White-tailed Sea Eagle should have become extinct in relatively few countries – the Faroes, Corsica, Sardinia, Austria, Denmark, Egypt and Israel, possibly also in Czechoslovakia, Hungary, Bulgaria and Syria. To this list must also be added Great Britain and Ireland (Fig. 16), where the species' demise has been better documented than most and to which the remainder of this chapter will be devoted.

From the account of Ussher & Warren (1900) and others, we know that at least 50 pairs once nested in Ireland; this is likely to be an underestimate since many of the more remote areas were rarely visited and undoubtedly held many more pairs than have ever been recorded. Placenames might provide additional clues. Inland pairs of Sea Eagles have occurred in the Comeragh mountains, Co. Waterford, where an eyrie is known to have been located in a high cliff overlooking a freshwater lough. Prior to 1831 three or

Distribution

Fig. 16. Previous distribution of the White-tailed Sea Eagle in Britain. Large dots are eyries known in the nineteenth century; small dots are eyries known prior to that time and open circles are some locations where a placename would suggest there to have been an ancient eyrie.

four pairs bred in the mountains of Mourne, Co. Down. There was a nest in Derry constructed on an island in the middle of a bog and it is interesting too that Sea Eagle bones were excavated from an eighth-century lake dwelling at Lagore, Co. Meath (Fisher, 1966a).

The remaining eyries in Ireland seem to have been coastal, a few of them on the east coast from Saltee, Co. Wexford, Lambay Island off Dublin and Fair Head, Co. Antrim. It was on the west of Ireland, however, that the species was most abundant. In Donegal, eyries were known on Malin Head, Torey and Owey Islands, Arranmore, Horn Head (two pairs) and at Teelin Head – it is likely that the name 'the eagle's nest' indicates the location of this latter eyrie. In Southwest Ireland in the 1870s a pair of Sea Eagles nested on the Blaskets where there is a place called 'Hollow of the eagles' on the island of Inishvickallaun; there is also a 'Mount Eagle' on the mainland nearby. Many other pairs must have frequented the spectacular cliffs of Kerry and a further four pairs are known from the neighbouring Co. Cork – Sheep's Head, Bere Island, Crow Island and the hills of Berehaven which were said to have been a great haunt prior to 1885. There is only one site known in Co. Clare; this was on the cliffs of Moher.

Some of the eyries in Galway were said to have been built in small trees on islets in freshwater lochs. A pair bred on Inishboffin to the south of which is a skerry known as 'eagle rock'. In all some 14 pairs are thought to have bred in the county, but it was the huge precipices of Mayo which, according to Ussher & Warren, 'long afforded a home to this our largest bird of prey'. Off Erris Head is an 'eagle island', while no less than four pairs were known from Achill. Single pairs frequented the cliffs at Loughmuriga, Alt More and the Spinks, while three other pairs nested on the great cliff of Alt Redmond (Jourdain, 1912). It was on the coast of north Mayo that the last pair of Irish Sea Eagles bred, probably in the year 1898.

It is curious that we have no written record of Sea Eagles' having nested in Wales, although fossils from the Pleistocene and later have been excavated from the Gower peninsula (Harrison, 1980). There are also countless references to the species in early Welsh literature. Dr W.G. Hale (personal communication) was once told of some climbers who visited an old Sea Eagle nest on the Llyn peninsula, near Nevin, in about 1880.

Yarrell (1871) mentioned that a pair had nested on the high cliffs of the Isle of Man until the eyrie was destroyed during a snow storm in 1818. According to Seebohm (1883), a pair also nested on Lundy, although an eyrie on Gannet Coombe seems to have been tenanted by Ospreys until 1858 (D'Urban & Matthew, 1892). A Sea Eagle was shot on passage on Lundy as recently as 1880 but it is likely that the species had ceased to breed there long before. It is also thought to have been Sea Eagles which nested on Dewerstone (Drewstaignton?) Rock on the River Plym in Devon (D'Urban & Matthew, 1892) probably some time during the eighteenth century or even earlier. Much of the ornithological literature on these English nests tends to confuse Ospreys and Sea Eagles but it is not unlikely that such eyries, especially the coastal ones, were in fact tenanted by the latter. A pair was recorded as having brought off two young from Culver Cliff on the Isle of Wight in 1780 (Yarrell, 1871) while a few other pairs lingered later in the Lake District. Here there is confusion with the Golden Eagle but it appears that at least one or two pairs of Sea Eagles nested in the Lakes until about 1794 (Yarrell, 1871), one of the eyries being described as occupied 'from time immemorial' (Macpherson, 1892). A nest on Wallow Crag near Haweswater was in use in 1787 while another pair simultaneously frequented Eskdale (Mitchell & Robson, 1976). Yarrell mentioned Keswick and Ullswater as being former nest sites and two Sea Eagles were seen at Ullswater as late as July 1835. This is often the date quoted in the literature for the last nesting by the species in England but there is no record of this pair actually having attempted to breed.

Within historic times Scotland seems to have been the stronghold of the Sea Eagle in Britain. I have traced more than 100 eyries, more than twice as many as in Ireland. Occasional pairs were said to breed in the south of the country, most of them inland; but the majority were to be found in the Hebrides, the western coasts and the northern isles of Orkney and Shetland. The lowland pairs ceased to breed at a comparatively early date last century.

Two cliff eyries were recorded from Wigtownshire – one on the Mull of Galloway and the other at Burrowhead near Whithorn; both ceased to be occupied by 1800. Saunders & Eagle Clarke (1927) claimed that Sea Eagles were still present in Galloway until 1836, while Bushnan (1834) is quoted in Baxter & Rintoul (1953) as finding 'not more than a dozen in Galloway from Nith to Mull'. Whether he was referring to nests or individuals is not clear. According to Gray (1871), Sea Eagles used to breed on Ailsa Craig, where there was a straggler shot in 1881 (Paton & Pike, 1929).

On 11th August 1812 *The Dumfries and Galloway Courier* informed its readership that 'on the island on one of the lochs on the farm of Stair, in the parish of Straiton and county of Ayr, the eagle had from time immemorial fixed his residence'. It is not surprising after such publicity that both eagles were trapped. The description apparently confirms their having been Sea Eagles and not Ospreys. Gray (1871) noted that Ernes were still breeding in inland districts of both Ayr and Dumbartonshire as late as 1840. Dresser (1871–1881) made reference to a pair in the parish of Hamilton, which may have been the pair nesting 'near Glasgow' and known to John Wolley (1902). The *Old statistical account* (Sinclair, 1791–99) described how the Eaglesham area possessed 'several woods especially on the banks of the river [which] together with the rocks in the neighbourhood are much frequented by eagles'. They would often perch on the holm or low ground where the village was later built and indeed, it is claimed, gave the village the name 'Eaglesholm' or 'Eaglesham'. This is disputed, however, by most authorities who seem to prefer a derivation from *Eccle fia-holm*, meaning 'the church in the hollow'. The lochs to the southwest of Eaglesham itself would certainly seem an ideal habitat for Sea Eagles while nearby runs a small river called 'Earn Water'. In this context 'earn' may mean white, as etymologists aver – as in the river and loch in Perthshire and indeed 'Erin' or Ireland. There sadly, the matter has to lie.

The inland areas of Kirkcudbright were once the haunt of Sea Eagles, and in 1767 both 'grey' and 'black' eagles (presumably White-tailed and Golden) were said to occur in plenty around the Merrick (Gray, 1871; Baxter & Rintoul, 1953). In Carrick Forest and Cairnsmore, Loch Skerrow is specifically mentioned by Sloan (1908) 'until a railway [now disused] arrived to disturb its shores'. The Sea Eagles which nested there apparently then moved to Glen Trool where 'seemingly annoyed by the continued improvements . . . they retreated to solitudes more profound' (Harper, 1876). The 'Eagle Island' in Loch Grannock perhaps harboured Ospreys although 'eagles' said to have nested on the crags above Loch Dungeon were more likely to have been Sea Eagles (since Dick (1916) recorded a battle there between an eagle and an otter): eventually the birds were driven from the area by local shepherds.

A Sea Eagle chick was removed from an eyrie around Cairnsmore of Fleet in 1858 and was held captive for many years at Cairnsmore House in Creetown (Maxwell, 1907; R. Roxburgh, personal communication). The Duchess of Bedford claimed that the species last bred in the area in 1852, although Maxwell puts it at about 1866 when a pair was trapped in the vicinity of Mullwharchar (Baxter & Rintoul, 1953; R. Roxburgh, personal communication).

The pair of birds reported nesting on Loch Urr in Dumfriesshire seem likely to have been Ospreys (Baxter & Rintoul, 1953). In Nithsdale, northeast of Closeburn, a hill called 'Earn Craig' may have been an ancient haunt, and at another crag of the same name on Criffel Sea Eagles were known to have bred until 1837 (Gladstone, 1910; Baxter & Rintoul, 1953). Apparently Sea Eagles also bred near Garwald Water in Eskdalemuir (Baxter & Rintoul, 1953). Sir Walter Scott immortalised a pair of eagles nesting on Loch Skeen where in his poem *Marmion* he wrote 'eagles scream from shore to shore'. They were almost certainly Sea Eagles rather than Ospreys and may have been the same pair which frequented the clefts of Talla Linnfoots in adjacent Peeblesshire. Every effort was made to extirpate them but in 1834, according to the *New statistical account* they reappeared and 'committed several depredations' on sheep. One of the Loch Skeen eagles was found dead on Moffat Water with its talons firmly embedded in the back of a salmon. Baxter & Rintoul (1953) record stray Sea Eagles on Loch Skeen in 1905 and 1908. Further east near Ettrickbridge there is a hill called 'Ern Cleugh'.

Leyden's mediocre verse quoted in Bolam (1912), suggests that Sea Eagles may once have bred in Roxburghshire, on

> Dark Ruberslaw, . . .
> where perches, grave and lone, the hooded Erne, . . .

It is interesting, too, that early in the seventh century St Cuthbert is said by Bede to have encountered an eagle in Roxburghshire. The bird had just caught a fish which the saint presumed to have been an offering to God, but typically the good saint left the eagle a half share. Placenames in Berwickshire and neighbouring counties hint at ancient inland haunts of the species – 'Earncleugh' hill and 'Earnscleugh's Water' in the Lammermuirs near Lauder (Muirhead, 1889), and 'Earnslaw' south of Duns. The cliff 'Earnsheugh' above the impressive seabird colonies of St Abb's Head was almost certainly a former nest site. James Fisher (1966a) was of the opinion that the description in the Anglo-Saxon poem *The Seafarer* (quoted at the head of Chapter 1) referred to the Bass Rock, and Wolley casually mentioned that this famous rock was indeed a former haunt of the Erne. It is tempting to reflect that this site could have been tenanted for nigh on 1000 years.

Further north still, in Kincardineshire, there lies another cliff called 'Earnsheugh' just to the south of the city of Aberdeen. The name 'Earnskillies' in Glen Clova may indicate an inland eyrie which Harvie-Brown (1906) reported to be occupied in 1813, although its exact location is now lost. Macgillivray (1886) believed that a pair of Sea Eagles once bred near Lochnagar (a mountain, despite its name!). Still in Aberdeenshire the *Old statistical account* noted how the species was 'now rare in the Forest of Burse'. Formerly eyries were located on Troup Head and on Pennan Head on the Buchan coast of the Moray Firth. The latter site is associated with an ancient prophesy of Thomas the Rhymer who, in the thirteenth century, is reputed to have said

> There should be an Eagle in the crags while
> there was a Baird in Auchmedden.

When the Baird family finally sold their Auchmedden estates some time later, the eagles ceased to breed on Pennan Head. They returned, however, as soon as the new owner married Miss Christina Baird. The estate passed out of the family once more and the eagles deserted the site for good. This tale seems to suggest one method whereby Sea Eagles could be encouraged to breed on our shores once more, if a suitable marriage could be arranged! Sim (1903) inferred that this story is more appropriate to Peregrine Falcons although Dr Ian Pennie (personal communication) knew of a similar story told of the Ernes nesting on Dunnet Head. Charles St. John (1893) located another east coast eyrie on the Soutars of Cromarty in Easter Ross. None of these North Sea coast sites could still have been occupied by the early nineteenth century and in view of the very tentative nature of the evidence for the existence of some of them, they may be very much older than that.

One or two localities in the Highlands also supported inland pairs of Sea Eagles. The flooded moors of Rannoch and Blackmount, for instance, were rich in freshwater lochs which at one time had much to attract Sea Eagles. In the 1830s a dozen or more might be seen sitting around on the stumps of old trees, while a local forester, Peter Robertson, recalled seeing 'packs of over twenty birds' at about this time (Booth, 1881–1887). Doubtless such gatherings included many juveniles or overwintering birds as only five nests are known in the area; one or two of these may have been alternative sites of the same pair. Loch Ba and Lochan na h-Achlaise, for instance, lie close to one another on either side of the modern A82 road from Glencoe. Loch Laidon, just across the border in Perthshire, also lies in this same catchment area. A fourth eyrie was found a few miles to the south on Loch Tulla. All of these nests were constructed in small trees on tiny islets on the lochs; the cliff eyrie above Loch Atriachan (possibly the site where two eggs were collected in 1866) was atypical of the area. Sea Eagles were also said to frequent the hills of Creran and Etive (*New statistical account*) but none of these nests in northern Argyll would appear to have survived beyond about 1870. Elsewhere in the county eyries were to be encountered on the cliffs of the Mull of Kintyre and Ardnamurchan Point. This latter site was deserted in about 1890, although a pair of adults was spotted there (from the mail steamer) in April 1913 and could well have been breeding (Beveridge, 1913).

On the mainland of Inverness-shire an isolated pair was known near Loch Laggan (Macgillivray, 1886) although there are also records of Ospreys breeding there (More, 1865). Another pair of Sea Eagles nested on one of the islands in Loch Loyne; the fir tree on which this eyrie was constructed was felled in 1835 but thereafter the eagles nested in a birch tree (Harvie-Brown & Buckley, 1895). Here also there may be some confusion with Ospreys which were said to have bred on Loch Loyne as late as 1916 (Sandeman, 1957b). Outside the breeding season Sea Eagles regularly turned up on the Glengarry Estates which at the time extended far to the west. In 1795 the *Old statistical account* noted three kinds of eagles occurring on the parish of Kilmallie, west of the Great Glen at Fort William – no doubt meaning Golden Eagle, White-tailed Sea Eagle and Osprey, the latter having bred on Loch Arkaig until 1908 (Sandeman, 1957b). Mrs J. Blackburn (1895) sketched a Sea Eagle eyrie at Inverailort, West Inverness-shire, on 20th June 1863 and this may have been the one depicted by R.R. McIan in his melodramatic painting entitled 'Robbing the eagle's nest' (Fig. 17).

There is little precise information available for the northwest mainland of Scotland but in Wester Ross Sea Eagles seem to have been scarce. Osgood Mackenzie (1928) mentioned their being rare around Gairloch compared

Fig. 17. 'Robbing the eagle's nest' – from a painting by R.R. McIan.

with the Golden Eagle, but he did recount how a keeper shot three adult eagles on Baos Bheinn, together with their two young in the eyrie – and all before breakfast! It is not clear which species was involved but the fact that there were two fledglings might suggest that they were Sea Eagles. Certainly a fine specimen of this species was trapped on Baos Bheinn in 1879 and was displayed in Inveran House (Dixon, 1886). There was also a well-known eyrie at Fionn Loch atop a crag on a small island known as 'Eilean na h-Iolaire'. When a boat was maintained on the loch the eagles were forced to move to Beinn Airidh Charr in the Fisherfield Forest. Two eggs were removed from this eyrie by Osgood Mackenzie's shepherd and it ceased to be occupied about 1850 (Dixon, 1886; Harvie-Brown & Macpherson, 1904; Mackenzie, 1928). Another eyrie was located on a sea cliff in Loch Broom (Harvie-Brown & Macpherson, 1904).

The Sea Eagle was more common in Sutherland and 'abounded' around Kildonan, Clyne and Edrachillis (Baxter & Rintoul, 1953). From 1848 to 1850 Wolley and William Dunbar robbed several eyries which Harvie-Brown (unpublished) identified as being in the Quinag, Ardvaar and 'Sal-en Geyers' areas. In *A vertebrate fauna of the North-west Highlands and Skye*,

Harvie-Brown & Macpherson mentioned a pair as nesting in the Ben Damph Forest until about 1889. Two eggs were taken by Wolley from Loch Stack in 1848, while another eyrie on Loch Fiag had become deserted about 3 years earlier. Charles St. John (1849) claimed that Ospreys bred on Loch Meadie but Wolley disputed this, the eyrie construction and some feathers he found nearby being in his opinion those of the Sea Eagle. A nest on Ben Hee, 10 miles inland, was situated midway between Lochs Fiag and Meadie but was in all probability that of a distinct pair.

The famous seabird cliffs on the island of Handa were said to have been vacated by Sea Eagles in about 1864 (Baxter & Rintoul, 1953) but in fact Gray (1871) noted a female that had been killed in 1867, while Colquhoun (1888) found eaglets in the eyrie in 1877. Two eyries were located on the high cliffs between Clo Mor and Garbh Island on the north coast of Sutherland and in May 1852 Wolley obtained two separate clutches on Whiten Head nearby; the two eyries were only $1\frac{1}{2}$ miles apart. An old nest on Island na Comb near Tongue is mentioned in the *Old statistical account*.

In 1868 Osborne (quoted in Baxter & Rintoul, 1953) considered that the Sea Eagle had been more common in Caithness than the Golden Eagle but only three eyries can be traced – perhaps the statement took into account the incidence of wandering immatures. The eyrie on Dunnet Head was a popular one with egg-collectors; it is known to have been robbed each year from 1847 to 1851, and was doubtless robbed in other years too. It was said that in 1840 an eyrie had been placed on the larger of the two stacks of Duncansby, and some time before 1887 a pair bred at the great seabird colonies on the Ord near Berriedale (Harvie-Brown & Buckley, 1895). This was perhaps the source of the two eggs collected 'near Dornoch' in April 1878 (Baxter & Rintoul, 1953).

The Sea Eagle's apparent scarcity amongst the southern Hebrides around Argyll may have been due to there being few sizeable seabird colonies, but it may also be true to say that these islands received fewer visits from naturalists; thus few eyries were recorded for posterity. A celebrated pair was known to nest on a cliff ledge in Catachol Glen in Arran until 1847. They returned in 1870 but deserted their clutch soon afterwards (Gray, 1871). The *Statistical accounts* of 1795 and 1846 refer to Sea Eagles building on the inaccessible precipices on Jura's west coast and this rugged island must have held several pairs. On Islay, eyries were known on the Mull of Oa and at Bolsa. Curiously, Baxter & Rintoul (1953) were told that nesting on Islay occurred even into the early years of this century. Colquhoun (1888) watched a pair with a young one on the island of Mull near Scallastle; pairs were also present at Burg on Loch Tuath and at Grulin (until 1871). Just offshore from Burg there is a skerry amongst the Treshnish Isles called

'Sgeir na h-Iolaire', whilst there is a spot on the Ross of Mull just opposite Iona which is still referred to as 'Nead na h-Iolaire' – the eagle's nest (M. Marquiss, personal communication).

To the north of Mull, the Small Isles – despite their restricted area – would seem to have been well endowed with Sea Eagles. A pair nested on Canna until they were robbed in 1875 (Baxter & Rintoul, 1953), while on the Isle of Eigg three pairs may have nested; it is however likely that an eyrie at the north end was merely the alternative nest of another pair whose last breeding attempt was made in 1886. About 9 years earlier a separate pair had ceased to nest on the spectacular Sgurr of Eigg (Harvie-Brown & Macpherson, 1904). Five pairs of 'eagles' ranged over the large and mountainous island of Rum but I would suspect that this properly referred to both species – possibly three pairs of Sea Eagles on the coast and two pairs of Golden Eagles on the inland crags. Dresser collected an egg – now in the British Museum (Natural History) – from the island but there is no date known. In the southeast corner is a rock wall called 'Sron na h-Iolaire' the nose of the eagle – which, being close to the main seabird colonies, would have been an attractive locality for Sea Eagles. Between the two World Wars this was used as a roost by Golden Eagles and in recent years two rusty gin traps have been found there – one of them being locked solid in a set position! Rum carries the distinction of being the only locality in Britain where a brood of three Sea Eagles has been recorded (Harvie-Brown & Macpherson, 1904). Lamentably they were all shot by the local shepherd, Duncan Macdonald of Skye, who completed a unique day's bag by shooting both adults as well. In 1866 the island's gamekeeper destroyed eight Sea Eagles in the one year, so it is not surprising that by the end of the century they were 'very rare' on Rum (Harvie-Brown & Buckley, 1892). A third site on Rum was occupied until about 1904, according to F. Cadogan (NCC Reserve records), but in 1907 a keeper took eggs from the nest and shot one of the adults. This was the last known breeding attempt on the island but the widowed bird must have quickly found a new mate for in 1909 both members of a pair were shot at the same locality (Rintoul diaries).

The Isle of Skye was, according to Gray (1871), 'the headquarters of this conspicuous eagle in the west of Scotland – the entire coastline of that magnificent country offering many attractions to a bird of its habits. Nearly all the bold headlands of Skye are frequented by at least one pair of Sea Eagles' (Fig. 18). Harvie-Brown & Macpherson (1904) knew of seven eyries between Waternish and Loch Bracadale – one of the nests being among 'the most picturesque eyries of the Erne on the west coast'. A further nine or 10 eyries were known to Gray (1871) from Gob na h-Oa (south of Loch Bracadale) to Loch Brittle. It is possible that some of these 17 nests were

Fig. 18. '. . . all the bold headlands . . .'

alternative sites only, but it is indisputable that the Sea Eagle was abundant on the west side of the island. Macpherson noted how as many as 40 might be attracted down to a carcass, while on one Skye estate no fewer than 52 eagles – nearly all of them Ernes – were shot or trapped within a period of 12 years (Gray, 1871). A few pairs were to be found on the east side of the island: at Storr (Seebohm, 1883), near Portree and doubtless also on Raasay where there is a hill named 'Beinn na h-Iolaire'.

Martin (1716) mentioned an eyrie on the huge basalt cliffs of the Shiants and this was still tenanted nearly 200 years later when, according to Eagle Clarke (Dr H.S. Blair, personal communication) a yachtsman, anxious to show off his prowess with a rifle, shot both adult eagles. About this time too, a clutch was taken from northeast Lewis (Dr J.W. Campbell, unpublished), perhaps from a 'Ness' pair visited in 1877 by Booth (Booth diaries). According to Macgillivray (1886) a pair annually reared young on the rugged crags beside Loch Suainavat, and this might be the 'Uig' site robbed in 1886 (Raeburn diaries). W.C. Taunton (Dr J.W. Campbell, unpublished) was shown a site on the cliffs in west Lewis in 1942 by a man who in his youth had taken an egg there. The Park area of Lewis was another stronghold of the species and many had been destroyed there last century by one local man, as many as 13 birds in one year (Harvie-Brown & Buckley, 1888). In 1877 Booth (according to his diaries) took two clutches there, while a third was taken further north in 1882 (Royal Scottish Museum).

Martin (1716) observed two sorts of eagles in Harris, one of them 'of a large size and grey in colour' – obviously the Sea Eagle. Gray (1871) knew several eyries on the island, one of them offshore on the Isle of Scalpay, while in 1868 Elwes (diaries) took a clutch from a cliff eyrie to the west of Tarbert. In 1877 Booth (diaries) raided a nest in a forest on the shores of Loch Seaforth.

Martin (1716) also knew of Sea Eagles in North Uist and eyries are documented on the hill of Lee and on a low cliff near Baigh Chaise, the latter deserted by 1880 (Baxter & Rintoul, 1953). Gray (1871) noted an eyrie on the island of Wiay and another on Benbecula. On South Uist a pair of Ernes nested where the lighthouse of Ushinish now stands. The eagles on Mount Hecla were thought by Gordon (1955) to have been Golden Eagles. C.M. MacVean had a pet Sea Eagle called 'Roneval', after another hill on South Uist from whence the bird came. The species was said to have bred on the islands to the south of Barra until 1869, and on Muldoanich there is a rock cleft called 'Slochd na h-Iolaire'. The species was said to breed on the stack of Arnamul at Aonaig cliff on Mingulay until about 1848 (Harvie-Brown & Buckley, 1888). There is an egg in the Parkin collection taken from south of Barra in 1863 (Baxter & Rintoul, 1953).

Martin (1716), again, mentioned Sea Eagles which had their nest on the north end of St. Kilda, and Steele-Elliot (1895) located the eyrie on the 530-m cliff of Connachair. The pair ceased to breed some time between 1829 and 1841 (Baxter & Rintoul, 1953) which is an early date for such a remote and suitable territory. But their disappearance is probably influenced by the skill of the St. Kildan people as cragsmen, who shared with the eagles a diet of seabirds; the remoteness of the archipelago would have made difficult recolonisation by eagles from the Hebrides over 40 miles away.

Of the Orkney Isles, Wallace (1693) wrote how 'Eagles or Earns are here in plenty'. In the seventeenth and eighteenth centuries, eyries were known near Houton Head (Gray, 1871), Mull Head (Low, 1813) and Costa Head (Macgillivray, 1886). Buckley & Harvie-Brown (1891) added that a pair formerly bred on Rousay, where there is still a spot known locally as 'Erne Tower' near Saviskaill Head. A pair was seen at Red Head on Eday in the spring of 1887 (Baxter & Rintoul, 1953) and there may once have been a nest site there. Macgillivray (1886) mentioned Sea Eagles breeding on South Ronaldsay, and Low (1813) on Switha. The sites on the low cliffs of Orkney must have been exceedingly vulnerable to human disturbance, which might explain their being deserted at a comparatively early date; probably the only sites to survive into the nineteenth century were on the high precipices of the island of Hoy. Buckley and Harvie-Brown (1891) knew of as many as 12 sites on this island but the actual number of breeding pairs must have been somewhat less than this. The famous sea stack 'The Old Man of

Hoy', a place called White Breast and an eyrie 2 miles inland near the Dwarfie Stane were all mentioned by Macgillivray (1886), and Gray (1871) added that two young were taken from a nest at West Craigs in June 1812. He also quoted Bullock's vivid description of these 'towering rocks rising to a perpendicular height of 1200 feet from the sea. About one-third of the way up this awful abyss a slender-pointed rock projected from the cliff, like the pinnacle of a Gothic building. On the extremity is a hollow, scarcely of sufficient size for the purpose [which] these birds had fixed on as a place of security for rearing young; the situation was such as almost to defy the power of man to molest their habitation . . .' But not quite, it transpired, for in 1853, 1866, 1868, 1877, 1899 and doubtless in other years between, clutches of eggs were stolen from the eyrie; in May 1888 a pair of young birds was removed from it (Buckley & Harvie-Brown, 1891). A surprising conclusion to the saga of the Sea Eagles of Orkney is that an egg was apparently collected there as recently as 1911 (*per* Jourdain Society).

The northernmost isles of Shetland might vie with Skye or the Hebrides for the most dense population of Sea Eagles in Scotland at one time. All the eyries were, according to Low (1813), on precipices, sea cliffs or stacks, except for one built on a low rock skerry in Moosa Water. In 1874 Saxby considered that there had been little diminution in numbers in Shetland despite their being less numerous elsewhere in Britain. A pair had, according to Eagle Clarke (1912), ceased to breed on Fair Isle between 1825 and 1840, the eyrie having been situated on Sheep Rock. The cliff still referred to today as 'Erne's Brae' may have been the territory of another pair but more probably was just an alternative site. Drosier (1831) noted an eyrie on the spectacular face of the Sneug on Foula. The islanders once protected the Great Skua (*Catharacta skua*) because they claimed it kept Ravens and eagles from their sheep. With over 3000 pairs of Bonxies now breeding on Foula, I am sure the Foula folk would gladly settle for a pair of eagles again! Clutches of Sea Eagle eggs were taken on Foula as recently as 1901 and 1902.

In 1845 a Free Church minister visited Fitful Head, the southernmost tip of Shetland (Anon., 1846) and romantically described a pair of eagles 'soaring through the azure depths of the air'. More prosaically, the older men of Dunrossness nearby observed to Venables & Venables (1955) that the Fitful eagles never touched lambs. Nonetheless with traditional but illogical practicality they repeatedly shot one of the pair. Its widow never failed to win a new mate until ultimately both were killed in the same winter; thereafter no more bred on Fitful Head although they are remembered locally by a rock in the Bay of Quendale nearby called 'Erne's Ward' – the eagle's lookout. An eyrie on the island of Noss was seemingly never occupied at the same time as the one across the Sound on Bressay – in all likelihood these were alternative sites of the same pair. The cliff called 'Erne's Hill' at Aith Ness in the north of Bressay probably belonged to another pair, distinct from the Bards of Bressay/Noss pair in the south. Wolley had an egg which had been taken on Noss in 1847, together with clutches taken on Bressay in 1856 and 1861. Seven eagles which were destroyed by shepherds on the tiny island of Vementry, had perhaps originated from a territory elsewhere. Venables & Venables (1955) noted how Ernes nested on Papa Stour, each year 'from time immemorial'. Across the water on Muckle Roe, part of the Shetland mainland, they were said to have been very abundant, with at least one pair suspected of having nested on 'Erne Stack' at the northwest end of the peninsula. A pair also had a nest at 'The Sail' in the north of Fetlar until 1885, but in the past the island may have held more than one pair (Evans & Buckley, 1899).

Certainly there were at least two resident pairs on Unst until about 1870. One bred at Hermaness and the other at Lund (Venables & Venables, 1955) where there is a rock called 'Erne's Hamar'. Several pairs may have frequented the island of Yell, for at the North Neaps there is an 'Eagle Stack' with an 'Erne's Stack' on the east side near Aywick. The only documented site on Yell however – at the Eigg on the West Neaps (sometimes called the Neaps of Graveland) – was the very last occupied nest in Shetland. The site had been known to Low (1813) as long ago as 1774 and doubtless remained in almost continuous use until 1904. In that year a single egg was stolen from it, eventually finding its way into the Royal Scottish Museum. It is recorded that the robber was caught and punished but in 1910 the eyrie was again raided by a collector 'locally reputed to have been an English clergyman'. Venables & Venables (1955) recorded that his climbing spikes were still to be seen in the rock. Tulloch (1904) observed that the female of this pair was an 'albino' and it was probably the same bird which had latterly frequented North Roe across Yell Sound. These sites may have been alternatives for the pair, Raeburn confirming how the North Roe pair had two or three different nest sites. Evans & Buckley (1899) added that two of the five pairs still remaining in Shetland towards the end of the nineteenth century 'not only change from one district to another but even from island to island'. Thus the confusing reports of an albino female occurring both on Yell and North Roe might be explained. In 1910 she was widowed but returned to her nest on North Roe 'to gaze out over the wide horizon and wait' (Fig. 19). No new mate came, however, for in 1916, far to the south in Skye, the last known breeding attempt had already been made. Two years later an old man shot this albino female, who seems to have been known and protected locally for nigh on 30 years (Lodge, 1946). Her sad death marked the final extirpation of the Sea Eagle in Britain.

Fig. 19. 'Albino' White-tailed Sea Eagle photographed in Shetland around 1912. This was the last individual known to survive in Britain and is probably the only British Sea Eagle ever photographed. It is thought to have been over 30 years old and was finally shot in 1918. (Photo: H.B. Macpherson, courtesy of Hon. D.N. Weir).

4 Breeding biology

> The White-tailed Eagle usually chooses for its retreat some lofty precipice overhanging the sea, and there in fancied security forms its nest and reposes at night.
>
> W. Macgillivray (1886)

As its name implies, the White-tailed Sea Eagle is a bird of the coast and Macgillivray's observation on its nest site is perfectly apt, but while this situation prevails in northern Europe and Greenland, further south Sea Eagles are to be found nesting inland where lakes and rivers prove ideal alternative habitats. Rarely do the birds nest far from water, be it salt or fresh. In East Germany, 75% of Sea Eagle eyries were situated within 3 km of a lake; one was 11 km distant but only 1 km from the shores of the Baltic (Oehme, 1961). Eighty per cent of these eyries were in forests, but the eagles exhibited a distinct preference for nesting on wooded islands or promontories, and nests were usually towards the edge of an open space – a clearing, marshy ground or even agricultural land. The few eyries in open habitat were never more than 150 m from forest (Oehme, 1961) and only one cliff nest was known, on the chalk cliffs of the Baltic island of Rügen. On this same coast two ground nests were used, one in 1880 and the other in the years 1904 to 1907 (Glutz von Blotzheim, Bauer & Bezzel, 1971). Ground eyries have been recorded elsewhere – usually on some remote offshore islet, on an isolated sandbank or in some secluded reedbed. One enterprising Norwegian pair chose to construct their nest on top of a 7 m-high seamark, on a coastal shipping route (Willgohs, 1961).

Despite such catholicism in situation, considerable discrimination may be shown in the final siting of the nest. Certain requirements have to be fulfilled. The eyrie must be easy of access for the eagles themselves and afford them a clear view of their surroundings. Yet it must provide adequate shelter from the elements and also security from predators and human interference. Through time the structure may achieve considerable proportions as more material is added; its foundations must be firm and strong. In the forests of Finland, one ornithologist has ventured to suggest that only one in every thousand trees may prove attractive to Sea Eagles.

Most nests are constructed in the crown of a tree, some 2–10 m from the top, usually against the trunk or in a large fork. Rarely will a nest be placed on a bare horizontal branch (Oehme, 1961). The height of an eyrie from the ground depends to a large extent on the tree species in which it is built. Five Romanian eyries built in Black Poplars and willows were from 15 to 25 m up; another placed in a fallen tree was only 7 m from the ground (Bannerman, 1956). On the lower Danube, Crown Prince Rudolf encountered an eyrie on 'a weak oak sapling' whereas six others were in high thick oaks, five in White Poplars, two in beeches and one in a Wild Pear tree (Bannerman, 1956). Glutz von Blotzheim, *et al* (1971) quoted 34 eyries in this region, ranging from 19 to 33 m above the ground. In East Germany, Oehme (1961) examined 177 nests: 65% of them were in mature pines, 22% in Copper Beech, 8% in oak and the remainder in alder (three nests), elm, poplar and birch (one nest in each). The eagles tended to choose the tallest trees and the eyries were placed from 8 to 30 m up (mean height 20 m). Schnurre (1956) found several nests at only 4–8 m. In Schleswig-Holstein (see Fig. 55) tall beech trees are invariably used with the eyries over 30 m from the ground (personal observation). Along the Baltic coasts of East Germany and in Sweden, pines are the most favoured (T. Neumann, personal communication; Banzhaf, 1937; Helander, 1975). Norway Spruce (Berg, 1923) and Larch (Dementiev & Gladkov, 1951) have been selected at high latitudes, where Flerov (1970) noted that either living or dead trees may be utilised.

Willgohs (1961) visited 98 eyries in Norway and only eight of them were in trees: two in pines, five in birch and one in willow. Martin Ball (personal communication) also noted one in an Aspen. This situation would appear to have prevailed also in Scotland at one time. John Wolley (1902) visited many eyries in his day but knew of only eight in trees. My own survey of the Scottish literature has revealed only seven: four in pines, and one each in alder, Rowan and birch. All the tree nests known to Peter Robertson, a stalker in the Rannoch area, were in pines (Alston, 1912). As in Norway the nests were rarely more than 7 or 8 m above the ground but often derived additional security by being located on small islands in freshwater lochs. A few of the Irish eyries were in similar situations; one was unusual in being constructed in a Yew tree (Ussher & Warren, 1900). On rare occasions some inland nests were placed on the ground such as one in the Rannoch area (Alston, 1912), another in Galway (Ussher & Warren, 1900) and a third in Moosa Water in Shetland (Low, 1813). In Shetland all but the latter eyrie were said to have been placed on precipices, stacks or seacliffs. In Scotland as a whole, 79% of eyries I have been able to trace in the literature were in cliffs; this compares with 86% of Willgohs' sample in Norway (Fig. 20). Among 22 eyries in Iceland which were described by Ingolfsson (1961), 17 (77%) were on cliff ledges and crags; the remainder were on low offshore skerries. In Greenland, 3% of 122 eyries were on offshore islets and skerries, the rest being on larger islands and on the mainland (Christensen, 1979); 70% were on cliff ledges and a further 16% on rock slopes, with 14% placed on the crest of knolls and cliffs, especially on the low flat-topped skerries. The latter eyries were often very exposed but most others were situated so as to derive maximum shelter from the severe climate. Eyries were often on south-facing slopes and at lower levels (from 21 to 60 m above the sea) than comparable nests in Norway. Also most of Christensen's Greenland nests were to be found in subsidiary fjords, these being more sheltered than the main fjords.

Throughout Greenland, Hansen (1979) noted densities of 0.3 to 0.6 occupied breeding territories per 100 km². In East Germany and east Ukraine, however, Fischer (1970) quoted densities of 1.6 to 2.0 pairs per 100 km². He also recorded five occupied nests on an 80 km stretch of the Yenesei River, and Willgohs (1961) noted nests to be no closer than 1–2 km apart. In areas with a deeply indented coast, territory size is difficult to measure precisely but in Norway it has been estimated at 600–900 hectares (ha) (Willgohs, 1961; Olsson, 1972). Fischer (1970) mentioned eight to nine pairs present on the island of Fugløy in Norway with its huge seabird colonies; in late summer as many as 75 eagles may congregate within its area of only 22 km²!

In calculating densities, the presence of several other unoccupied nests belonging to the same pair may cloud the issue. Most pairs of Sea Eagles possess two or three alternative sites. Willgohs (1961) gives a mean of 2.5 per pair, with one pair having as many as 11 alternatives. These tended to be in similar situations and only four pairs possessed both cliff and tree eyries. In Scotland, at least two pairs were recorded as having such contrasting sites; the eagles nesting on the flat island in Fionn Loch, for instance, moved to an alternative eyrie on a nearby crag (Harvie-Brown & Macpherson, 1904). A failed breeding attempt may stimulate a pair to

Fig. 20. A 6–7-week-old eaglet in a typical coastal eyrie in northern Norway, June, 1975. (Photo: J.A. Love).

abandon one nest in favour of another but this is by no means always the case. Some repeatedly unproductive pairs will persist at the same eyrie for several years, while other pairs shift site for no apparent reason.

An unfortunate feature displayed by many coastal sites is their accessability to man. This obtained even in parts of Greenland where persecution from sheep farmers was persistent and heavy. Christensen (1979) identified 62% of eyries as 'easily accessible' and only 13% foiled all his attempts to reach them. Such a situation prevailed also in Iceland, Norway and in Scotland where few eyries seemed to escape the persistent efforts of a determined egg-collector. John Wolley for instance described one nest in Sutherland on a rocky bank overgrown with small trees: it lay 'in a sort of great chair of rock, was perfectly accessible from any direction right or left, above or below, and a man could get within a yard or two from above without in the least disturbing the bird'.

In spite of their accessibility to man many eyries enjoyed a prolonged history of occupation. Local people in the Highlands of Scotland were often heard to comment how a particular eyrie had been in use 'since time immemorial'. The celebrated eyrie on the Shiant Islands in the Outer Hebrides was mentioned by Martin Martin in 1716 when it may already have been long tenanted. Harvie-Brown & Buckley referred to it again in 1888, saying of the pair then in residence: 'Long may they continue in their inaccessible retreat; and may the broken overhanging columns . . . resist the wear and tear of time, and prove a sheltering roof to them'. Barely two decades later two eagles were shot there (Dr H. Blair, personal communication) and the site was abandoned after a tenancy lasting two centuries. Similarly an eyrie in Iceland remained in use for about 150 years. In complete contrast, two pairs of the Darss peninsula in East Germany used a new nest in each of 6 years! (Glutz von Blotzheim *et al.*, 1971).

Very often an eyrie proved to be located in an extremely picturesque situation – worth all the effort in reaching it, as John Wolley (1902) testified: 'To enjoy the beauties of a wild coast to perfection let me recommend to any man to seat himself in an Eagle's nest'. And whilst doing just this somewhere in the northwest of Scotland, he penned this vivid account of a typical construction. The eyrie was placed

> on a sort of triangular ledge, a small Rowan tree touching it in front. The rock is scarcely overhanging. The nest is made chiefly of dead heather stalks, with a few sticks for the foundation, the largest of which are above an inch in diameter and two feet long. It is lined with a considerable depth of moss, fern, grass and *Luzula* . . . The hollow is small for the size of bird, and very well defined. There is a

rank sort of smell, but no animal's remains in or near it: several feet below me is an old nest.

As there may be flexibility in the choice of situation, so is there variability in nest construction. On secure rock ledges there may be little need for much nest material and often a good thickness of guano-like soil scooped up into a shallow depression will suffice. On a bare rock shelf, heather and juniper stems, stipes of dried *Laminaria* or sticks up to 1.5 m long and several centimetres thick may be incorporated; indeed any mixture of whatever may be available. If there is a scarcity of material, such as on some remote skerries, seaweed, driftwood, even glass or metal floats may be collected, and once or twice even the dried skeleton of a sheep (Willgohs, 1961).

The actual shape of the structure tends to be determined by the ledge. Oval or angular nests are most common (Christensen, 1979) with an average diameter of about 1.5 m. Two nests lying side by side can sometimes amalgamate into one huge platform. The largest nests tend to be those constructed in trees – rigid, strong structures circular in shape, up to 2 m in diameter and 3 m deep (Fig. 21). Glutz von Blotzheim *et al.* (1971) quoted one weighing 600 kg. These large nests tend to be accumulations of many seasons' repairs and additions. Willgohs (1961) described a Norwegian eyrie of about 8 years' standing, having some 2 m depth of material. The following year a further 0.3 m was added, although the nest was not used that year. By next season it had been built up to nearly 3 m. A nest on Loch Ba in Argyll presented 'a most extraordinary appearance when viewed from a little distance' (Gray, 1871), while Colquhoun (1888) added how it comprised 'not less than a cartload of sticks and twigs'. An Irish eyrie constructed in a yew tree (mentioned earlier) measured 3 m across (Ussher & Warren, 1900) whilst a similar one in Greece reputedly served as a refuge for a notorious local bandit! (Bannerman, 1956). Another in Norway reached twice the height of a man until eventually the tree could support it no longer and broke (Willgohs 1961). Such an outcome was not uncommon particularly with added weight from winter snows or during stormy weather.

The shallow bowl of the nest which will contain the eggs measures some 0.2–0.45 m across and about 0.1 m deep (Willgohs, 1961). It is lined with lichen, moss, seaweed, ferns, grass, woodrush (*Luzula*), heather (*Calluna*) or *Empetrum* – indeed whatever suitable material may be to hand. In many Greenland eyries (Christensen, 1979) and in some Scottish ones, sheep's wool may be employed.

In captivity it is usually the male who initiates nest-building, by bringing in tufts of grass (Fentzloff, 1977): Fischer (1970) claimed that it was the male who selected a site for the nest and was responsible for the early stages.

Breeding biology 53

Fig. 21. White-tailed Sea Eagle eyrie over 3 m tall
with annual accumulations of nest material, Norway.
(Photo: J.F. Willgohs)

Thereafter both sexes are involved, branches being collected from the ground but never broken off trees. Activity seems to be most intense during the morning, and latterly it may be the male who undertakes the greater share of the work, the female merely rearranging material on the nest platform. The whole construction may be completed within 18 days, although fresh greenery can be added throughout the breeding cycle. Fentzloff (1977) noted, however, that his captive pairs restricted the bringing of fresh pine sprays to one or two days prior to laying and, again, around the time of hatching. This habit has generated much speculation as to its function – the provision of insulation or humidity for incubation, or to cover faeces and carnage on the nest, aiding sanitation once the young have hatched.

Sea Eagles have been known to utilise old nests of other species such as Black Kite, Buzzard or Raven, while pairs have been seen to evict Ospreys and Red Kites. In return Imperial Eagles (*Aquila heliaca*), Saker (*Falco cherrug*) and Peregrines have been seen to use old eagle eyries. On separate occasions both Peregrine and Saker actually succeeded in evicting pairs of Sea Eagles from their eyries (Fischer, 1970; Glutz von Blotzheim, et al., 1971). An interesting case from Mecklenburg is documented by Deppe (1972); a pair of Sea Eagles had occupied an eyrie tree for about 30 years until timber operations forced them to move. Two years later the eagles returned to find their original nest had been taken over by Peregrines. Undeterred they constructed a new eyrie higher up the tree. Many aerial battles ensued but the falcons failed to expel the eagles, which only gained respite from the attacks once they had alighted on their own nest. This uneasy situation persisted throughout that year (1947) and the next when both eagles and Peregrines each succeeded in rearing one youngster. Willgohs (personal communication) has known Sea Eagles, Peregrines and Ravens to nest one above the other on the same cliff face. Contests between Golden Eagles and Sea Eagles are not infrequent, but of this more will be said in a later chapter.

There are reports of Great Horned Owls (*Bubo virginianus*) nesting within occupied nests of Bald Eagles, while one pair even succeeded in evicting the eagles altogether (Broley, 1947). Crown Prince Rudolf encountered several Sea Eagle eyries in Hungary harbouring whole colonies of Tree Sparrows (*Passer montanus*) (Bannerman, 1956). Pied Wagtail (*Motacilla alba*), Treecreepers (*Certhia familiaris* and *C. brachydactyla*), Redstart (*Phoenicurus phoenicurus*), Crested Tit (*Parus cristatus*), Starling (*Sturnus vulgaris*) and Stock Dove (*Columba oenas*) have all been recorded nesting within old Sea Eagle eyries (Fischer, 1970). Fischer also knew of Buzzards and kites nesting in the vicinity; together with Ravens they would help themselves to prey remains on the eagle eyrie. He also referred to one small wood in Macedonia which contained one pair of Sea Eagles, two pairs of Imperial Eagles, no less than 12 pairs of Black Kites and a heronry!

The Sea Eagle will undertake some nest building as early as December, although courtship may occur even earlier. Fischer (1970) recorded display in the autumn but considered it to be amongst pairs which had failed in their breeding attempt that season. Many pairs remain faithful to their breeding territory throughout the year but in Arctic regions some adults may be forced south for the winter, when the lakes freeze. Well-established pairs seem to remain together over winter so that nest-building and repair, and courtship can commence as soon as they return to the breeding grounds. In Kandalaksha this occurs around early March, but not until mid-April further north (Flerov, 1970).

Sea Eagles are gregarious creatures, immatures, sub-adults and some migrant adults congregating together at favourable feeding grounds or to roost. On territory some adults may be tolerant of neighbours while others will defend their nest site jealously. Intruding adults, especially a lone male, will sometimes provoke fierce fighting. Banzhaf (1937) found a male lying dead with his throat lacerated; he lay 15 m from the nest but the fight had obviously taken place there, for the clutch was destroyed. The same author had witnessed three other fights to the death, but such cases are rare. Fischer (1970) watched an adult male engage and lock talons with a younger male, while the female of the pair flew around calling excitedly. The young bird was eventually seen off by the male and only then did his mate participate. Early ornithologists delighted in recounting dramatic fight scenes. Gray (1871) described four Ernes soaring together yelping loudly while two others fought. Locked in combat, the two fell to the ground in a cloud of feathers, one being so injured that it was unable to rise. Another pair fighting above Loch Lomond were less fortunate: 'Both at last became so firmly grappled to each other by their talons that they were precipitated into the water. The uppermost regained the power of his wings, but the other was taken alive by a Highlander who witnessed the scene' (McWilliam, 1936). What became of the luckless bird is not divulged although an Orkney gamekeeper, capturing both combatants in like manner, 'had them stuffed' (Buckley & Harvie-Brown, 1891).

Shepherds on Skye told Harvie-Brown & Macpherson (1904) that the Sea Eagle 'paired in early spring; at this season it may also be seen to indulge in deadly conflict with other members of its fraternity'. Sometimes a widowed female may treat a new male as hostile, but on the whole it seems that replacement mates are accepted with a period of days only. On the other hand there are instances of reluctant lone males who remain unpaired for many years although single females may occasionally enter their territory (Fischer, 1970).

Talon-grappling (Fig. 22) has been recorded in several raptor species – Kestrel (*Falco tinnunculus*), Sparrowhawk, Red Kite, Hen Harrier (*Circus cyaneus*), Buzzard and Golden Eagle – where it seems to be a component of territorial defence; it usually involves members of the same sex (M. Marquiss, personal communication). Among *Haliaeetus* species, however it seems to be an integral part of courtship also. (A type of talon-grappling also occurs when young eagles are begging from flying adults and strive to snatch food from their talons.)

Preliminary courtship may be seen in October and November but becomes most intense in early spring, especially around the immediate vicinity of the nest. The male or sometimes the female, may perch near the

Fig. 22. White-tailed Sea Eagles talon-grappling.

nest, calling excitedly with head thrown back (Fig. 23) in what Cramp and Simmons (1980) call 'sky pointing'; I prefer the term 'long call' since it is an intensely vocal display and rather reminiscent of one used by gulls, etc. In full voice, Fish Eagles may touch their back with the head, but I have never seen Sea Eagles achieve this. The wings are held partially open at the shoulders with the tips touching over the tail. This vocal display achieves maximum effect in flight when a pair may soar around together, usually in the vicinity of the nest, calling loudly in duet. Either partner may lead, from 1 to 6 m ahead, or they may instead soar in opposing circles (Glutz von Blotzheim, et al., 1971). One may swoop upon the other who responds by tilting to one side with hardly a flap of a wing, or else it may roll over to touch talons momentarily before the pair separate. This may be repeated or else gain in intensity as talon-grappling or 'mutual cart-wheeling' when the pair tumble out of the sky with feet locked firmly together, to disengage sometimes only a few feet from the ground. This exciting display, so well

Breeding biology

Fig. 23. 'Long call.'

developed in the genus *Haliaeetus*, is evocatively described for Bald Eagles by Walt Whitman (1819–1892) in his poem *Dalliance of the eagles*:

> Skirting the river road, (my forenoon walk, my rest)
> Skyward in air a sudden muffled sound, the dalliance of the eagles,
> The rushing amorous contact high in space together,
> The clinching interlocking claws, a living, fierce, gyrating wheel,
> Four beating wings, two beaks, a swirling mass of tight grappling,
> In tumbling, turning, clustering loops, straight downward falling,
> Till over the river pois'd, the twain yet one, a moment's lull,
> A motionless still balance in the air, then parting, talons loosing
> Upward again on slow-firm pinions slanting, their separate
>   diverse flight,
> She hers, he his, pursuing.

During the period of courtship the pair will remain together much of the time, often perching or roosting side by side on the eyrie. This usually heralds coition, which may be initiated by either male or female. Gerrard, Wiemeyer & Gerrard (1979) described in captive Bald Eagles how the female might move towards the male with head lowered and wings slightly open, uttering a soft, high-pitched call. The male responds by calling, flapping his wings and moving his tail up and down. If it is the male who approaches first, the female bows low with wings apart and invites him to mount her.

White-tailed Sea Eagles behave similarly (Glutz von Blotzheim, *et al.*, 1971), the female crouching low, almost flat, on the ground, a low perch or on the nest; sometimes both members of the pair will assume this posture simultaneously. The head and neck are outstretched, the wings half-open and the tail held level with the rest of the body (Fig. 24). The male then flies or jumps on to the female's back calling loudly and flapping both wings to maintain his balance. After about 12 seconds he dismounts and both may then sit quietly for several minutes. Copulation has been recorded to occur four or five times in the space of an hour and a half. According to Willgohs (1961) this may be at any time of day but Fentzloff's captive pairs tended to copulate in the afternoons only (Fentzloff, 1977). Gerrard *et al.* (1979) found that 44% of the observed copulations amongst their captive Bald Eagles took place between the hours of 0700 and 0900, the rest being scattered throughout the day. The act of coition is usually undertaken close to the eyrie although Willgohs (1961) has witnessed it up to 3 km away. In both Germany and around the White Sea, White-tailed Sea Eagles have been observed hotly engaged in sexual activity on the cool ice of frozen lakes! Two copulations witnessed in the wintering grounds may have been an established pair wintering together (Glutz von Blotzheim, *et al.*, 1971).

Display flights end rather abruptly once the eggs have been laid. Weather seems to influence the date of laying the first egg remarkably little. Some females show a surprising degree of consistency in this respect, with the net result that average laying dates in any population may differ by only a week or two from one season to the next. The date of laying does vary considerably with latitude, however. The White-tailed Sea Eagle breeds through some 35° of latitude and may breed up to 3 months or more earlier in the south than in the north. In Israel and Iraq laying commences during January; in Greece and the lower Volga, during February; in Germany, during March; and in the Arctic, in late April to early May (Table 3). Willgohs (1961), referring to John Wolley's accounts, suggested that laying in Scotland was about 3 weeks later than in Norway, but within the great latitudinal range of the latter country there is much variation. Eggs may be laid from late March to early April, sometimes into mid-May in Finnmark, but even near the Arctic Circle new-laid eggs have appeared as early as 5th March. In Britain, Seebohm (1883) concluded that clutches were to be found from early March with laying continuing until mid-April. Jourdain (in Bent, 1961) was more specific, giving extreme dates of 15th March and 28th April, with most laid during the middle of April. But Jourdain's figures are dates on which eggs have been found in nests and not actual laying dates. I know of two clutches from the Outer Hebrides which were collected as early as 4th and 18th March, with five others actually laid around 11th,

Breeding biology

Fig. 24. Female (right) soliciting.

17th, 20th and 21st April. It would seem therefore that Sea Eagles would have begun to breed in Scotland at about the same time as Norwegian birds do nowadays which, considering Scotland's lower latitude, is a few weeks later than one might expect.

It is difficult to make a strict comparison in breeding times of the Sea Eagle and the Golden Eagle in Scotland. Brown (1976a) stated that Golden Eagles in Scotland commenced laying from 1st March and continued through to mid-April; most clutches were completed by 25th March. MacNally (1977) found that most pairs in mid-Inverness-shire laid their eggs during the latter half of March. My own experience on Rum, further west, is that only about 50% of the observed clutches had appeared by the first week in April. The earliest clutch taken by collectors was dated 18th March. Thus it would seem that both Golden Eagles and Sea Eagles began to breed in Scotland at roughly equivalent times, if anything the Golden Eagle being marginally earlier.

Of the White-tailed Sea Eagle, the Rev. Morris (1851) was moved to comment that 'the eggs . . . by a merciful provision are few in number'. Banzhaf (1937) considered that young females laying for the first time tended to lay only one egg. Certainly single egg clutches are not uncommon. Two is the normal clutch, however (Fig. 25), less frequently three. None of the 14 clutches known from Galway in Ireland last century contained more than two eggs (Ussher & Warren, 1900), although they recorded a three-egg clutch from Co. Kerry in 1869. In the Outer Hebrides, Macgillivray (1886) had never encountered three eggs, but Elwes (diaries)

Table 3. Latitudinal variation in laying dates of White-tailed Sea Eagle

Latitude	Country	Laying period	Source
c.35°N	Iraq	Late January, early February	Jourdain (in Bent, 1961)
	Israel	January	Y. Yom-Tov (personal communication)
c.40°N	Turkey	Mid-January to mid-February	Jourdain (in Bent, 1961)
	Greece	January to March	Cochrane (in Bannerman, 1956)
	USSR (Volga)	End of February	Dementiev & Gladkov, 1951
c.50°N	USSR (Ukraine)	March	Kozlova (in Bannerman, 1956)
55°N	West Germany	March (especially first half)	Neumann (personal communication)
	Germany	End of February to April	Niethammer, 1938
57°N	Scotland	15th March–28th April, especially mid-April	Jourdain (in Bent, 1961)
55–60°N	Sweden	Late March, mostly early April	Helander, 1975
61–71°N	Norway	Early march to mid-May (especially late March and early April)	
66°N	Iceland	April/early May	Willgohs, 1961
66°N	Greenland	Late March to mid-May (especially late April)	Ingolfsson, 1961
67°N	USSR (White Sea)	Late April to early May	Salomonsen, 1950
			Flerov, 1970

Breeding biology 61

Fig. 25. Sea Eagle nest containing a clutch of two eggs, Norway. (Photo: J.F. Willgohs).

knew of several such clutches and actually collected one in 1868. Harvie-Brown & Macpherson (1904) described clutches of three as 'not uncommon' and were told that Duncan Macdonald on Skye had once found four eggs in a Sea Eagle eyrie. Other four-egg clutches have been recorded from Poland (Fischer, 1970), Romania (Witherby, Jourdain, Ticehurst & Tucker, 1943) and Norway (Willgohs, 1961). In 1975, Johan Willgohs and I visited an eyrie in northern Norway which we had been told by a local fisherman had contained four eggs; the huge nest collapsed out of its tree before incubation was completed. There is always the possibility that an infertile clutch might survive intact over the winter to be added to the following season, or even that two females might lay in the same nest; such an event has never however been recorded for the Sea Eagle. We must assume therefore that some females may be capable of laying such an abnormally large clutch. Captive pairs are known to lay additional eggs that same season, should their first clutch be removed.

Only one of the 55 clutches which I have traced as having been collected in Britain contained three eggs, most contained two but 45% were single eggs. It is possible that some of these may have derived from larger clutches, having been split or partially destroyed once collected. The average British clutch from this sample is calculated at 1.56 eggs. A further 10 clutches are described in the literature, two of them containing three eggs and one (mentioned above) four. These average 2.10 eggs per clutch but it should be remembered that the literature will contain a bias towards unusually large clutches. Only two of Willgoh's sample of 58 Norwegian clutches were derived from the literature, the remainder were from museum records (28 clutches), the author's own observations (7) or from his correspondents (21). They provide a mean of 2.16 eggs per clutch, significantly larger than the entire British sample of 65 clutches (Table 4). There is a paucity of similar data from other countries although in the White Sea area Flerov (1970) recorded two clutches of one egg, four of two and one of three eggs (mean 1.86 eggs). Glutz von Blotzheim, et al., (1971) quoted a mean clutch of 2.6 eggs from Kazakhstan but the sample size is not mentioned. Particular females are often consistent over several years in the number of eggs which they can lay. Willgohs (1961) knew of one who laid three eggs each season from 1957 to 1960, and this has also been noted in the Golden Eagle (in British collections, and M. Marquiss, personal communication).

Once the first egg has been laid there is a lapse of 2 or 3 days before the next. Fiedler (1970) once suspected an interval of a week in his captive pair at Vienna Zoo, but he was never able to confirm this. In Tel-Aviv Zoo, over six seasons a female Sea Eagle exhibited a 2-day lapse once, 3 days on three occasions and once 4 days between the first and second eggs; intervals of 4 and 7 days were recorded between the second and third eggs. (Prof. Mendelssohn, personal communication). In 1972 Claus Fentzloff's captive eagle, Clara, laid her first and only egg of the season; the two eggs in each of her subsequent three annual clutches were laid at intervals of 2 days. In her fifth and sixth years of breeding, the interval became 3 days. A second female, Thora, had a 3-day lapse between eggs in one year and 4

Table 4. *Clutch size of the White-tailed Sea Eagle in Britain and Norway*

	Source	No. of eggs per clutch				Mean ± Standard Deviation (Sample size)
		1	2	3	4	
Britain	Literature	3	4	2	1	2.10 ± 0.99 (10)
	Collections	25	29	1	–	1.56 ± 0.53 (55)
	Both	28	33	3	1	1.64 ± 0.64 (65)
Norway	Willgohs, 1961	5	40	12	1	2.16 ± 0.59 (58)

days the next (Fentzloff, 1977). A third female from Eekholt Wildpark (Brühl, personal communication) showed 2 days between eggs in one year. There is little information available upon this and other aspects of egg-laying in wild populations; perhaps fortuitously, for at this time eagles are very susceptible to disturbance at the nest.

Glutz von Blotzheim, *et al.*, (1971) recorded fresh egg weights of 138–145 g. The second egg is invariably smaller than the first; Fentzloff (1977) recorded a mean of 124 g from five second eggs, compared with 135 g from four first eggs. If a replacement clutch was laid, he noted the egg could be smaller still – one weighed only 112 g. A second clutch may be produced if the first is destroyed or experimentally removed early in incubation. The intervals necessary to effect this were noted by Fentzloff to be 19 days in one clutch and 29 days in another. Bannerman (1956) claimed that early clutches may be susceptible to frost and snow and if addled could be replaced within 3 weeks or so. Fentzloff's results would seem to confirm this. The Eekholt female incubated an infertile egg for 58 days before laying a replacement almost immediately; this egg subsequently hatched.

The eggs of the White-tailed Sea Eagle can be described as blunt-ovate to almost round in shape, rough in texture and white in colour. Some appear glossy, and all tend to become stained yellowish when incubated for some time. In a sample of 51 eggs from Norway and 54 eggs from Britain (which Willgohs reckoned to be similar in size) the length varied from 65 to 89 mm and the breadth from 53 to 64 mm (Willgohs, 1961; Witherby, *et al.*, 1943). These tend to be slightly larger than eggs from Germany and Denmark, for example, and a sample of 12 eggs from Asia Minor were smaller still. The largest eggs would appear to be laid by *H.a. groenlandicus* (Table 5). Thus, since egg size might be related to the size of the female, this may reflect a cline in body size of the species, increasing from south to north.

As in all eagles, incubation begins as soon as the first egg is laid, with the female undertaking the greatest share of incubation (Fig. 26). Willgohs

Table 5. *Geographical variation in dimensions of White-tailed Sea Eagle eggs*

Country	Average length × breadth (mm × mm)	Sample size	Source
Turkey	71.8 × 56.2	12	Jourdain (in Bent, 1961)
USSR (Balkans)	72.7 × 53.5	?	Fentzloff (personal communication)
USSR (south)	73.3 × 57.9	11	Cramp & Simmons, 1980
Germany/Poland	74.8 × 58.3	24	Glutz von Blotzheim et al., 1971
Denmark	74.5 × 56.9	21	Schiöler, 1931
Scotland	75.8 × 58.7	54	Witherby et al., 1943
Norway	75.7 × 58.8	51	Willgohs, 1961
Iceland	74.7 × 56.4	11	Gudmundsson, 1967
Greenland	75.3 × 58.3	29	Schiöler, 1931
	77.5 × 58.1	21	Jourdain (in Bent, 1961)
	78.0 × 59.3	23	Hartert (in Fischer, 1970)

Fig. 26. Female incubating.

(1961) estimated that the male's contribution amounted to 27% of the total period. The female invariably sat throughout the night, with the male relieving her early in the morning and again for brief spells during the day. The number of change-overs averaged about six per day. If a relief failed to take place on the nest itself, it was usually because the male had departed before the female had returned. One of Fentzloff's captive male eagles, Korsar, was unusually attentive to the eggs and often excluded the female for prolonged periods. The longest time for which Willgohs observed the eggs to be unguarded was 20 minutes, although on one occasion, with one parent standing guard nearby, they were uncovered for 48 minutes. In total, the eggs were exposed for no more than 4% of the incubation period. This compared with 2% in a pair of captive Bald Eagles (Gerrard et al., 1979). The attentiveness of this incubating pair was related to temperature and the eggs were rarely left uncovered in cold or windy conditions when the risk of chilling was highest.

Fentzloff noticed how the adult eagles moved around the nest cup with their massive talons firmly and safely clenched; the hind claw directed forward and shielded by the other toes so as not to puncture the eggs. As a bird settled to incubate it would shuffle from side to side supported on its slightly open wings, pivoting on its beak, which it would hook on some convenient branch in front. I once watched one of Fentzloff's females, Clara, actually alter the position of a stick so that she could more conveniently hook her beak over it.

Willgohs (1961) noted the male to bring food to an incubating female (Fig. 27) on four occasions, and on three of them she left the nest to eat before resuming her duties. He estimated that an egg took 38 days to hatch. This has since been confirmed by Fentzloff, whether the egg had been under the female or incubated artificially. Fischer (1970) gives a range of 34 to 46 days (most 38–42 days) but it is not clear whether this refers to the whole clutch or individual eggs. Mendelssohn (personal communication) has recorded 38, 39, 39, 40 and 40 days for five first eggs, but 35 and 37 days for two second eggs: overall the mean incubation period for the seven eggs is calculated to be 38.3 (± 1.8) days.

About 24 hours after the first cracks appear in the shell the scrawny youngster emerges. The duty parent then alters position so that the nestling can lie under the breast feathers while the remaining egg or eggs fits (fit) into the brood patch as before. Since incubation had begun with the first egg, subsequent eggs hatch 2 or 3 days apart (Fig. 28). The female is especially reluctant to leave the nest at this time, and Fischer (1970) stated that in some pairs only the female brooded the young and that while doing so she was fed by the male. Once the chicks are slightly older, however, the male also broods them, but not at night. From this time, change-overs start to take place away from the nest (Willgohs, 1961). When the male appears with food the female adopts a begging posture not unlike that of the older chicks. She then tears up the food (Fig. 29), swallowing the larger portions herself and presenting the smallest pieces to the chicks. If the female is absent when the male arrives at the nest, he may take it upon himself to feed the young. A captive male of Fentzloff's, Korsar undertook 30% of feeds and Willgohs found that wild males contributed as much as half; but on the whole the male's efforts are of shorter duration than the female's. Feeding bouts will last from a few to 22 minutes and occur at intervals from 1 to $4\frac{1}{2}$

Fig. 27. Adult White-tailed Sea Eagle landing on eyrie with a char (*Salvelinus alpinus*); its mate awaits on the eyrie, Greenland. (Photo: F. Wille).

hours apart. A chick may receive food as many as 11 times in 24 hours and these may take place at any time of day. At one of Willgohs' study nests, feeding of the chick occurred first in the early morning from about 0530 and sometimes again at mid-day, but most often in the afternoon and evening until about 2130.

As the chicks become older their food requirements increase and bird prey often takes the place of fish as the favoured prey. Surplus food lying in the nest also then becomes less frequent, as the eaglets can tear up prey for themselves. Willgohs noted that the adults removed waste food only twice during his prolonged observations at the nest.

The newly hatched chick weighs from 90 to 100 g. It is covered in thick

Breeding biology

Fig. 28. Sea Eagle chick a few days old with unhatched egg, Norway. (Photo: J.F. Willgohs).

Fig. 29. Adult Sea Eagle tearing up prey for 6-week-old chick, Norway. (Photo: J.F. Willgohs).

creamy-white down which is longest on the head. The down on the head, chin and throat is more white, but become greyer behind and under the eyes, on the wings and on the rump. There is a bare patch on each side of the belly; the back of the tarsus and the lower part of the legs and feet are also devoid of feathers. The legs and cere are pink, the beak black. Tiny chicks are brooded by the parents almost continuously. About the eighth day they may be uncovered for periods of up to 30 minutes but the adult remains on guard at the nest rim. The chicks can crawl around feebly and utter a weak cheeping if disturbed. The parent eagle will offer tiny pieces of food, turning its head from right to left until the chick is stimulated to accept. As the parent does this its tail also moves accordingly and such characteristic behaviour can be an early indication to a distant observer that the chick has emerged from its shell. Much saliva is produced by the parent during such feeding bouts and this may aid the chick in digesting the food, as Fentzloff (personal communication) has discovered in vultures. At first, only four to eight morsels may be offered each hour, but by the sixth day this has increased to 20 or more larger pieces every 2 hours. At the age of 10 days or so (Fig. 30), the chick may commence pulling food from the parent's bill.

By the twentieth day (Fig. 31) the initial growth of white down is replaced by a thick woolly coat of longer, coarse grey down. This tends to be darker on the crown, underparts and flanks. The legs and cere now appear pale yellow. The chick is more ready to snatch food from the parent but still does not pick it up from the floor of the nest. It remains uncovered for longer periods and from about 4 weeks of age stands weakly on the tarsi, or moves around the nest more with the aid of its forelimbs. Thus it is able to defaecate clear of the nest lining. The voice is stronger and more varied, and some chicks achieve a weak form of threat behaviour if disturbed; otherwise they lie flat and still. By about 30 days the first feathers, tipped with white fluff, begin to emerge through the down.

By the sixth week (about 40 days) the eaglet (Fig. 32) is more firm on its feet and will feed voraciously from the parent as well as snatch food from the floor of the nest. It will even make the first attempts to feed itself, but it cannot yet use its talons. It is able to swallow more skin, bones and feathers. Patches of grey down remain on the thighs, under the wings and on the throat. Flecks of down remain obvious, tipping the feathers of the nape, but these are finally lost by the seventh week. The eaglet is by now more alert, will pick up sticks or food around the nest and defaecate well over the edge. It now attempts its first clumsy wing exercises. After the eighth week (Fig. 33) only the long feathers of the wings and tail have yet to develop fully. The eaglet stands on its feet for long periods, stretches, preens and exercises vigorously with much jumping and running around. By 9 weeks it is able to

Breeding biology

Fig. 30. Eaglets 12 and 14 days old.

Fig. 31. Eaglet at 3 weeks of age.

Fig. 32. Adult Sea Eagle at nest with 5- or 6-week-old chick, Norway. (Photo: J.F. Willgohs).

Fig. 33. Eaglets at about 8 to 9 weeks of age.

hold food firmly in both feet. The youngster is fully developed at 10 weeks and most will undertake their first flight at about this time; occasionally though an eaglet might be found to remain in the eyrie for an additional week or two.

The brood normally numbers either one or two young (Table 6). Both Morris (1851) and Saxby (1874) claimed that any third egg in a clutch is invariably addled; but this is not always the case. There exist from the British Isles several records of three young in nests. Elwes was told of one in

Table 6. *Brood size of White-tailed Sea Eagles*

Country	No. of chicks in brood								Source
	1		2		3		4		
	No.	%	No.	%	No.	%	No.	%	
Iceland	7	78	2	22	–	–	–	–	Arnold & Maclaren, 1938
Norway	39	42	49	53	4	4	1	1	Willgohs, 1961
	54	47	59	52	1	1	–	–	Norderhaug, 1977
Britain	6	40	7	47	2	13	–	–	Love (unpublished)
West Germany	12	71	5	29	–	–	–	–	Neumann (personal communication
East Germany	9	50	9	50	–	–	–	–	Fischer, 1970
	25	96	1	4	–	–	–	–	Dornbusch, 1977
Poland	13	28	33	70	1	2	–	–	Banzhaf, 1937

the Outer Hebrides (Baxter & Rintoul, 1953) and in 1876 three young were also taken from an eyrie in Co. Kerry (Ussher & Warren, 1900). When Duncan Macdonald was employed as a shepherd on the Isle of Rum in the 1820s or 1830s he shot three young out of an eyrie, together with both parent birds; but this was the only instance of a brood of three in his wide experience. Such accounts find special mention in the literature and must inflate any estimate of average brood size calculated for the British situation last century (Table 7). Elsewhere in Europe fewer than 4% of broods contained three young; Willgohs (1961) recorded four instances in Norway and another is mentioned from Poland by Banzhaf (1937). There is also a unique observation of a brood of four chicks in Norway (Willgohs, 1961).

About 33% of nesting attempts, sometimes as many as 75%, fail to produce any young. The success of a pair may vary from year to year. Fischer (1970) cited several examples of this; between the years 1946 and 1955, one pair on the Danube reared five broods with only one young and one brood of two. A pair of Sea Eagles in Germany reared two young 4 years in succession while another nest succeeded in producing three young in each of 2 successive years. Under normal circumstances, broods of one and two occur with approximately equal frequency although the latter would appear to have been especially common in Poland between the wars. The Icelandic situation might have been influenced by persecution which was rife at that time, whilst it is known that brood sizes in Germany are severely affected by pesticides. Thus in Germany, as in Finland and Sweden, the brood size per successful nest is less than that pertaining during the pre-

Table 7. *Average brood sizes (per successful nest) in White-tailed Sea Eagles*

Country	Mean no. of chicks per brood	No. of broods	Source
Greenland	1.6	45	Hansen, 1979
Iceland	1.4	47	Ingolfsson, 1961
Norway	1.6	160	Willgohs, 1961; Norderhaug, 1977
Britain	1.7	15	Love (unpublished)
Poland	1.7	47	Banzhaf, 1937
Poland (1950–69)	1.4	18	Bogucki, 1977
East Germany	1.1	27	Dornbusch, 1977
West Germany (1948–50)	1.1	14	Fischer, 1970
West Germany	1.3	17	Neumann (personal communication)
Sweden (Lappland)	1.4	66	Helander, 1975–79
Sweden (Baltic)	1.3	171	Helander, 1975–79
Finland	1.3	174	Stjernberg, 1981
USSR (Estonia)	1.2	98	Randla & Oun, 1980
USSR (Lower Kazakhstan)	1.9	?	Gratschew (in Glutz von Blotzheim *et al.*, 1971)

pesticide era, or that found in unpolluted areas such as Norway and Greenland. This will be discussed at greater length in Chapter 7.

Breeding failure can be brought about by many agencies. Human persecution, of course, remains to this day an important influence. Pesticides can result in embryo deaths while eggs can be infertile, accidentally damaged or lost to a predator. Both eggs and young can be lost during periods of inclement weather while whole nests can be destroyed in strong gales. It may be impossible for an investigator to deduce the ultimate cause of breeding failure: had Willgohs (1961) not witnessed a tiny chick killed beneath a heavy branch in the nest structure, its cause of death might never have been known. A variety of predators may raid Sea Eagle nests. Flerov (1970) mentioned martens and bears, while Fischer (1970) added crows, Ravens and a Marsh Harrier (*Circus aeruginosus*); even foxes and Wild Boar have been known to eat young which have fallen from the nest.

Amongst some birds of prey, sibling aggression can reduce brood size (see Newton, 1979). Fischer (1970) held that young White-tailed Sea Eagles lived peaceably together in the nest. Willgohs (1961) concluded sibling aggression to be rare in this species, although he noted some minor incidents. Fentzloff (1977) observed the older chick in one of his captive broods to peck at its younger sib, but if the latter remained quiet until its precocious partner was replete, it would duly receive its share from the parent. Such a situation tended to prevail amongst two 4-week-old chicks which I hand-reared on Rum. It was significant, however, that the younger chick was more willing to pick up scraps as its mate was fed. While the older chick long expected me to present it with food, the youngster early demonstrated its ability to tear up food for itself.

Meyberg (1978) encountered cases of deliberate aggression among Sea Eagles in Romania. He quoted Moll who claimed that broods in East Germany are regularly reduced thus. Arnold & MacLaren (1938) noted how, from about 3 weeks of age, the elder chick in one Icelandic nest quickly gained in weight at the expense of its younger sib. One observed attack lasted 15 minutes until the small, weaker chick ultimately died. On June 1st Flerov (1970) found two well-grown but still down-covered chicks in an eyrie in the White Sea area; by 17th June only one remained alive and the other, very emaciated, was found trampled into the nest. A pair of White-tailed Sea Eagles which bred in an aviary in Vienna Zoo in 1961 laid two eggs, both of which hatched, but one chick disappeared a few days later and in each subsequent season only one chick fledged (in some years however one of the eggs had been infertile). In 1969 Fiedler removed the first egg as soon as it was laid and replaced it once the second had appeared; thus they hatched simultaneously and both chicks, being of comparable size, were reared successfully. This was an early instance, if not the first, of manipulative techniques being employed to increase the reproductive output of a Sea Eagle pair. Another instance was carried out by Helander (1979) in a wild Swedish nest in 1978. The eyrie contained three young, two weighing 4.2 and 3.3 kg, respectively, and the third only 1.9 kg. This 'runt' had been pecked about the head and so was removed and hand-reared for several days before being introduced into another nest; this contained two eaglets about a week younger but of comparable weight to the foster chick. All three fledged successfully. Harold Misund (personal communication) encountered a brood in northern Norway where the third chick was a weakling. Once removed and hand-fed it quickly improved in condition and became one of four chicks to be transported to Rum that year as part of the reintroduction project. Otherwise it would certainly have died in the nest.

Thus, instances of what Seton Gordon (1955) has termed 'Cain and Abel rivalry' cannot be as uncommon among Sea Eagles as was once thought, but it is true to say that compared with Golden Eagles in Scotland, two young are more often reared to fledging.

In Central European nests, large, well-grown young can be found as early as the end of April, while further north, in Norway, most young will fledge about mid- or late July (Willgohs, 1961), and about 3 weeks later in the

White Sea area of the USSR (Flerov, 1970). It seems that the situation in Scotland was similar to that in Norway and Wolley (1902) for instance, recorded two young in late July which were 'well able to fly'; they crouched 'side by side with their necks stretched out and chins on the ground, like young fawns, their frightened eyes proving that they had no intention of showing flight'. (The eaglets were removed and taken into captivity.) Harvie-Brown & Macpherson (1904) recount a curious incident when a young lad robbed a cliff eyrie in Skye; one of the eaglets immediately took flight but fell into the sea below. An adult then swooped down and picked up the eaglet in its talons but was unable to carry it off. Herrick (1924) detailed a similar tale of a Bald Eagle which tried to lift an injured eagle from the water. Doubtless, behaviourists can postulate simple explanations for such acts but one tale told by Saxby (1874) can safely be dismissed as fiction. An eyrie was found in Shetland which contained two young and an addled egg. The egg was collected and when a man returned a few days later for the young he claimed that the parent eagles had removed both nest and young to a more secure site on the other side of a rock cleft! One would suspect that something other than the eagle's egg had been addled!

After fledging, the young eagles may remain with their parents for several months, gradually acquiring hunting skills and learning to fend for themselves. Observations on this stage of the life cycle are sparse but Wille (personal communication) believes that on leaving the nest the young will be 100% dependent upon the adults for at least 4 weeks.

It will be fully 5 years before the young attain full sexual maturity and during this time they are distinctly gregarious. In Norway, immatures converge upon suitable feeding grounds for the winter, such as Lofoten, where as many as 30 or 40 may roost together. A roost is usually in trees on steep slopes of the offshore islands (Willgohs, 1961). At dawn and dusk the eagles fly regularly to and from the feeding areas and even in the continuous darkness of an Arctic winter they manage to find food. The roosts also remain in use throughout the bright nights of summer (Willgohs, 1961).

Recoveries of ringed birds indicate that few Norwegian Sea Eagles travel any considerable distance. One was found 720 km south near Karlstad in Sweden (Hagen, 1976) and another on the Waddenzee in Holland, while four chicks colour-marked in Norway have been sighted in southern Sweden (Helander, 1980). In contrast, young from Finland and Sweden tend to disperse in a southwesterly direction to the shores of the Baltic (Helander, 1975, 1980; Stjernberg, 1981). One has been recovered 520 km west in northern Norway and another as far south as Bulgaria, nearly 2000 km away (Fig. 34). Some immatures reared in West German nests

Breeding biology

Fig. 34. Ringing recoveries of White-tailed Sea Eagles in Europe (data from Helander, 1975; Glutz von Blotzheim *et al*, 1971; Saurola, 1981). There is no evidence to show that Icelandic Sea Eagles leave that country. Up to 1976 a total of 264 chicks had been ringed in Norway (eight recovered), 138 in Sweden (21 recovered), 79 in USSR (nine recovered), 54 in Finland (10 recovered) and 35 in West Germany (six recovered). Dotted lines are birds ringed as adults.

may remain for the winter on the Baltic coast, but one has been reported in Holland, and two others sighted on the Danube near the Austrian/Hungarian border (Fig. 35*d*). Similarly, two chicks ringed in East Germany were later recovered only 50 km from their natal area, but others have been found in the Gironde, France (1520 km southwest), Belgium (220 km westsouthwest) and Italy (1030 km south), the latter two from the

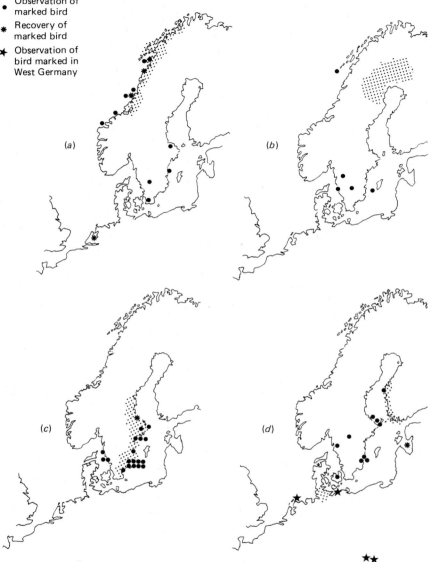

Fig. 35. Sightings and recoveries of colour-marked White-tailed Sea Eagles (after Helander, 1980; Saurola, 1981). Stippling indicates area where marked. (a) Norway: 221 ringed; 10 sightings; 5 recoveries. (b) Lappland: 25 ringed; 5 sightings. (c) Swedish Baltic Coast: 42 ringed; 18 sightings; 1 recovery. (d) Finnish Baltic Coast: 20 ringed; 10 sightings; 1 recovery. West Germany: 12 ringed; 4 sightings.

same brood! (Glutz von Blotzheim et al., 1971). From the vicinity of the White Sea, movement south begins in September and is completed by November; in mild winters some adults may remain behind (Flerov, 1970). The return movement takes place during February and March. Recoveries of ringed birds indicate that these birds may have wintered as far south as Hungary and Italy, over 2000 km away. Counts at one Baltic wintering area, on the River Elbe in East Germany, revealed the pattern of build-up amongst these wintering eagles. Combining records for 37 seasons showed that arrival began in November; numbers reached a peak in January and declined again during March and early April (Glutz von Blotzheim et al., 1971).

In the course of their return movement, occasional birds may overshoot their natal area. One Romanian immature was recovered 330 km north of its ringing site. Another ringed as a chick near Odessa on the Black Sea was caught and released again the following April near Kharkov, 510 km to the northeast, before being found dead 5 months later, some 240 km to the south (Glutz von Blotzheim et al., 1971). This is reminiscent of the Bald Eagle situation in Florida where locally bred young move north with overwintering eagles returning to northern USA and Canada (Broley, 1947).

A southerly dispersal seems once to have prevailed in Britain. Most of the Sea Eagles found wintering in the south of the country were presumably locally bred rather than immigrants from the continent. Among 102 records I have traced, 80% occurred during the months October to March. Nearly all the birds were immatures and most were ultimately shot. Ringing demonstrates the high mortality suffered by young eagles. Amongst 16 Swedish recoveries, seven had died during their first winter and an eighth at 1 year old (Helander, personal communication). A further 29 unringed birds have been reported dead in Sweden – 14 were juveniles and five aged from 1 to 4 years (Helander, 1975). In Norway, five of nine recoveries had died within their first year of life (Hagen, 1976). Cramp & Simmons (1980) analysed 90 recoveries of Sea Eagles ringed in Europe from 1911 to 1969; 92% had died before reaching maturity and 70% of these in their first year. In recent years White-tailed Sea Eagles have been ringed in Greenland by Danish ornithologists; 10 have been recovered thus far, all having been shot within their first year of life (Hansen, 1979).

Persecution seems to be a major problem in Greenland as it is in East Germany where Oehme (1977) examined 194 Sea Eagles found dead or injured between the years 1946 and 1972: 39% had been shot; accidents accounted for a further 6%, power cables being a major hazard; territorial disputes contributed another 7.5%, and 13% of the birds had died from poisoning (it is likely that this was an underestimate). Ninety-five of

Oehme's sample were adult birds, a surprisingly high proportion since Cramp & Simmons (1980) estimated adult survival to be about 70% per year. The oldest known ringed bird was 27 years old when it was found dead (Helander, 1975). In captivity, however, Sea Eagles can live for well over 40 years (Lodge, 1946). A captive Sea Eagle taken in Kirkcudbright in 1858 died in May 1900 at an age of 42 years; the bird was by then totally blind (Maxwell, 1907). There are countless tales, no doubt all apocryphal, of eagles attaining the magical age of 100 years. No less an authority than Seton Gordon (1955) repeated an unlikely story of a Golden Eagle reputedly shot in France in 1848; around its neck was a gold collar bearing the date 1750 but the ornament could easily have been passed on to several different eagles in the intervening 98 years. In captivity, Golden Eagles have been known to live up to 60 years (Hancock, 1973).

In summary, White-tailed Sea Eagles occupy a wide range of aquatic habitats and within these may choose to nest in a variety of different situations – trees are typical of freshwater areas and cliff ledges of coastal ones. The nests, of sticks and branches, can attain immense proportions as they are added to each year. Within any territory there may be two or three alternative eyries, sometimes as many as 11. A normal clutch contains two large, white eggs, laid in January in the south of the species' range or as late as May in the high Arctic. The eggs are incubated for 38 days, and 10 weeks after hatching the eaglets are ready to leave the nest. Often two young are reared in one nest but instances of the older chick's killing its younger sibling are not unknown. Both parents participate in incubation and in the care of the brood. The young attain sexual maturity at about 5 years of age, until which time they tend to be gregarious and may disperse to areas of abundant food; this dispersal seems to be in a southwesterly direction, although it is not a regular migration in the strict sense of the word. Most will die in their first year and only 10% may reach adulthood. Sea Eagles are comparatively long-lived, even in the wild where a maximum age of 27 years is known from ringing recoveries. Diet and hunting techniques will be discussed in the following chapter.

5 Food habits

> A sea eagle exists – it has very sharp vision and while hovering in the sky, once it sees a fish in the sea plunges headlong on to it, and splitting open the water snatches it to its breast.
>
> Pliny (AD 23–79)

It is possible that Pliny could have been referring to the Osprey rather than the Sea Eagle. Both can catch fish but the Osprey with its long flexible wings can hover as it searches for its prey. Its fishing ability is further enhanced by its slit-like nostrils being closed as the bird submerges and its plumage being dense and oily to improve its waterproofing. Other features, though less well developed, are to be found also in the White-tailed Sea Eagle. Horny spicules cover the soles of the feet and, together with the long sharp talons, facilitate the grasping of slippery fish. The legs and feet are sturdy and strong to withstand the impact of the strike. I have also noticed Sea Eagles reverse the outer toe to hold a fish, so that two toes are held forward and two back, but this is accomplished without the facility of the Osprey. Rarely does the Sea Eagle plunge-dive under water, preferring instead to snatch fish lying close to the surface or in shallow water: thus the spectrum of prey species available is more restricted. The Sea Eagle is of course a larger bird; its powerful beak (Fig. 36) is a conspicuous feature, used to pierce and tear with a strong twisting and pulling action. Yet it is wonderful to witness the dexterity and finesse with which a parent eagle will employ this formidable tool to present even the tiniest morsel of food to a newly hatched chick.

The fishing ability of the White-tailed Sea Eagle was much admired by the

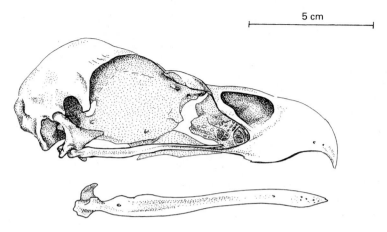

White-tailed Sea Eagle

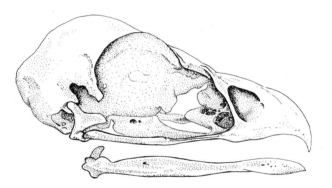

Golden Eagle

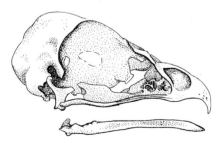

Buzzard

Fig. 36. Skulls of Golden Eagle and Buzzard to compare with that of the White-tailed Sea Eagle. All are male birds and are drawn to scale.

fisherfolk of Shetland who, in their ignorance of functional morphology, attributed this proficiency to the supernatural and religiously annointed their own baits with eagle fat in the firm expectation of improved catches. They also believed that as soon as an Erne appeared overhead fish would rise to the surface, belly upward, in a hopeless gesture of submission (Pennant, 1774).

The Sea Eagle shares with other raptorial birds a highly developed sense of sight. Despite the 'eagle eye' being proverbially the epitome of visual acuity, it has attracted surprisingly little study. Walls (1942) noted how, in the eye of a raptor, the sensory cells tend to be concentrated in the upper hemisphere of the retina so that, as an eagle flies, images from below can be perceived more clearly. When perched and wishing to engage an object on the same plane, however, an eagle might risk all dignity by turning its head completely upside down, focussing with a penetrating but distinctly quizzical stare. In all birds the eyes are large in relation to the head, and those of the Sea Eagle exceed 30 mm in diameter – larger than those of a human (Fig. 37).

Although sight is of prime importance to the Sea Eagle in the location and pursuit of its prey, it can be attracted to carrion by the mere presence of other eagles, vultures or corvids. There is no evidence that smell is involved. It has recently been shown, however, that New World Vultures such as *Cathartes aura* utilise smell to detect carcasses (Stager, 1964). With rank carrion, the Sea Eagle may find it a blessing to lack any olfactory powers!

As it soars in its search for food, the long, broad wings 'fingered' at their tips, present the familiar eagle silhouette. With air turbulence – and hence drag – most pronounced at the wing tips, these deeply notched and emarginated primaries function as aerofoils: thus drag is reduced, the wings effectively lengthened and lift enhanced. The total wing span of a White-tailed Sea Eagle may exceed 2.5 m.

The crop or oesophagus of birds has an elastic wall and that of the Sea Eagle can hold several days' sustenance, permitting it to exploit temporary abundances of food. Thus it can afford prolonged stints loafing on a favourite perch or soaring effortlessly in air thermals. A brief radiotelemetry trial on Rum in 1977 indicated that one female spent only 8% of its time in the air (Love, unpublished) although during the 4-day observation period the weather was bad and may have grounded the bird more than normal; also, being newly liberated it was probably somewhat reluctant to venture far. Nonetheless the results were similar to more detailed studies by Brown (1976*b*, 1980) on the closely related African Fish Eagle which spent some 75–90% of its day perched.

Fig. 37. Head of juvenile female, Croyla, about 4 months old, Rum, August 1980. (Photo: J.A. Love).

Such extended periods of inactivity, so typical of predators, together with its large bulk in relation to body surface, serve to minimise the Sea Eagle's food requirements. Yet the bird does display a healthy appetite. I was able to determine the daily food intake of seven fully grown birds (aged 3–6 months) which were temporary captives on Rum as part of the reintroduction project (Love, 1979). In the first trial, lasting 3 months, three females were fed a variety of food types – birds, fish, venison and goat meat. Two years later, in 1977, two other females and two males were fed entirely upon mackerel for a period of 3 weeks. The actual amounts consumed each day varied from nil (usually following a big meal the previous day) to 1.4 kg. The average daily intake was much the same for all the females (571 g net weight) and for both males (456 g) – about 9% of body weight (Table 8). Although food was always provided in excess, the birds were weighed only at the conclusion of the trials so it is not known whether they had increased in weight in the meantime. Being retained in individual cages, 4 m square and 2 m tall, opportunities for exercise were limited. Craighead & Craighead (1969) demonstrated an inverse relation between body size and food consumption among a range of raptor species. Thus one would expect the

Table 8. *Daily food intake of captive White-tailed Sea Eagles on Rum (Love, 1979)*

Sex	Bird no.	Food intake per day Mean ± standard deviation (g)	Range (g)	No. of days	Body weight (kg)	Daily food intake as % of body weight
Female	1	553.2 ± 272.0	0–1221	106	6.9	8.0
	2	577.2 ± 281.0	156–1221	102	6.4	9.0
	3	584.0 ± 297.0	57–1448	74	5.9	9.9
	4	617.5 ± 191.3	201–1027	20	5.6	11.0
	5	543.5 ± 221.4	261–1031	20	5.5	9.7
Male	6	502.4 ± 178.3	168–820	20	5.0	10.1
	7	408.6 ± 139.2	222–681	20	4.8	8.5
TOTALS						
Females		570.7 ± 290.4	0–1448	322	6.1	9.4
Males		455.5 ± 164.9	168–820	40	4.9	9.3

smaller Golden Eagle to require more food relative to its body weight than the Sea Eagle. However, a Golden Eagle studied by Fevold & Craighead (1958) consumed only 5–6% of its body weight per day. This eagle was kept in slightly different circumstances so the comparison may not be valid; it was tethered but allowed frequent bouts of exercise while being fed entirely on venison (which, in addition, has a lower moisture content than fish). It is surprising also that the male Sea Eagles, being smaller than the females, consumed an equivalent amount of food (relative to their body weight) to that consumed by the females. The series of measurements made on the males was considerably smaller, however. Amongst the females there was an indication that the largest bird ate less, proportionate to her body weight, than her smaller peers (Table 8). The results obtained from the Rum study confirmed observations made upon Sea Eagles elsewhere. For instance, Uttendorfer (1939) had concluded that a full-grown captive bird, maintaining constant body weight, required about 500 g of meat per day – about 10% of its body weight. In a wild situation, sustained activity imposes greater food demands but by how much is not easy to determine. One study by Kirkwood (1980) on Kestrels (*Falco tinnunculus*) indicated that to fly free this small, very active species required two or three times more food than it did in captivity. Willgohs (1961) was able to observe that a pair of wild White-tailed Sea Eagles, together with their three fledglings, each consumed about 625 g of food per day. This is much less than Kirkwood's results would predict, but was considered to have been a slight underestimate since during the 8-day observation period all three eaglets lost body weight.

Captive Sea Eagles on Rum could consume two whole mackerel, each weighing about 350 g, within 6 or 7 minutes. Seven or eight small fish were swallowed whole within a space of only 2 minutes. In each instance the crop was filled to the size of a small grapefruit. Meat is more time-consuming to devour, requiring 20 minutes or more before an eagle is satisfied (Fig. 38). Willgohs (1961) estimated that one wild eagle had eaten 2 kg of fish in a single meal while a 3-year-old, free-flying male on Rum consumed 1.8 kg – about 40% of his own body weight.

The actual weight of prey eaten is, of course, less than that actually killed, due to wastage. Most of the feathers, large bones and often the beak, feet and sternum of birds are discarded, together with the gills, tail, etc., of larger fish. Indigestible material such as small bones, feathers or fur which are ingested accidentally are later regurgitated in a pellet (Fig. 39). Normally one pellet may be produced every 2 days or so but the frequency is very much determined by diet. Pellets appear most regularly when an eagle is feeding upon bird prey. These pellets also tend to be the longest and bulkiest (Table 9) since they often contain the head and beak, or a complete leg wrapped up in miscellaneous bones and feathers. On a diet of small fish, tiny pellets (about 40 × 24 mm), consisting entirely of grass, may be produced. Such vegetation may be ingested deliberately or accidentally while the bird is tearing up its prey. The grass may serve to cushion the bird's digestive tract against sharp fish bones. I have watched captive eaglets even tearing at and ingesting newspaper, presumably towards this same end.

Pellets collected in quantity from beneath a nest or a favoured roost can be employed to investigate diet. Biases can however be introduced. Already we have seen how easily an eagle can digest many fish bones so that they may rarely appear in pellets. On the other hand feathers, fur and animal bones are the stuff of most pellets and can readily be identified. Not every prey species contributes to a pellet with the same degree of facility however. Mick Marquiss was able to study this problem in Goshawks (*Accipiter gentilis*) by comparing the prey remains located at plucking posts with the content of pellets regurgitated by the hawks; he has kindly made his findings available to me. Fur-bearing prey, especially lagomorphs and squirrels, tended to be detected more easily in pellets than at plucking posts, whereas game birds, which are relatively loose-feathered and easily plucked, are underestimated in pellets. Buzzards, although less useful for such studies, yielded broadly similar results: small mammals had a higher frequency of detection in pellets compared with plucks. Carrion tended to leave bone chips and encouraged the appearance of grass in pellets; afterbirths, frequently found at Buzzard nests, left no trace at all in pellets.

Recent studies of Sea Eagles in Greenland by Danish ornithologists have

Table 9. *Dimensions (mm) of 66 pellets produced by three captive female White-tailed Sea Eagles, Isle of Rum, 1975*

Length: 80.7 ± 27.9 (mean ± standard deviation) 34–157 (range)
Breadth: 31.0 ± 5.6 (mean ± standard deviation) 20–45 (range)
Longest: **157** × 27   Shortest: **34** × 22
Widest: 114 × **45**   Narrowest: 39 × **20**

Fig. 38. Juvenile tearing up prey.

Fig. 39. Immature regurgitating a pellet.

further elucidated the problems in data collection (Kampp & Wille, 1979; Wille, 1979). Small fragments of prey and pellets collected from nests and brought back to the laboratory for identification showed a distinct bias towards bird prey (68% of items) at the expense of fish (20%). In contrast, more complete items lying in a nest and readily identifiable *in situ*, comprised 58% fish and only 30% birds. A completely different pattern emerged when Wille (1979) mounted prolonged surveillance at five different nests (over three breeding seasons). He used automatic cameras to photograph each food item carried in by the adults. Fish was found to constitute 93% of the diet and birds only 5%. Prey remains collected at 40 other eyries in the same district consisted, as expected, of 68% fish and 16% birds (see Fig. 40).

The use of such sophisticated equipment is both expensive and time-consuming. Normally one has to rely on the simple identification of prey remains while bearing in mind its limitations, for example, that fish will be underestimated and birds overestimated. While such a technique will not reveal the precise diet of the predator in question, it can be used to compare diets from different habitats (Table 10). Sea Eagle prey lists now exist from several countries. As we would now expect, most show a preponderance of bird prey (39–87% of items), although one exception is Fischer's sample of 79% fish from the middle reaches of the Danube where bird prey was in short supply. Only in the inland Lappland reserve on the Kolsky Peninsula of northern USSR (Flerov, 1970) did mammal prey assume any importance (41% of 172 prey items); here birds and possibly fish too, might be less abundant than in coastal areas. Flerov's data also indicated how Sea Eagle diets may vary within a relatively restricted area and this was confirmed in Greenland. Kampp & Wille (1979) found that in the south of the country seabirds were scarce and more fish were taken. Even within the same small area differences in diet may be marked; fish constituted up to 50% of the items taken well up the Greenland fjords, while on the outer coasts and skerries the numerous seabird colonies provided an alternative abundance of prey. Individual pairs of eagles too may display dietary preferences, dependent upon what is most available within their hunting range.

It should also be borne in mind that prey found in a nest may not reflect what the adults themselves consume. Fish with easily digestible bones might be given preferentially to eaglets, since their growing bones and feathers will demand an enhanced calcium intake. Meat is rich in phosphorus and therefore less desirable at this time since it can seriously disrupt the ratio of calcium to phosphorus within the eaglet's body: studies on vultures feeding exclusively upon carrion (Mundy & Ledger, 1976) have demonstrated how bone fragments in meat regurgitated by the parents are essential to the chick's normal development.

Food habits

Table 10. *Diets of White-tailed Sea Eagle*

Country	No. of items	% fish	% birds	% mammals	Source
Greenland	557	34	58	8	Kampp & Wille, 1979
USSR (Kandalaksha)	523	12	75	13	Flerov, 1970
USSR (Kolsky)	172	20	39	41	Flerov, 1970
Finland	?	31	55	11	Stjernberg, 1981
Norway	1612	34	57	8	Willgohs, 1961 (and unpublished)
Sweden	478	49	47	4	Helander, 1975
East Germany	384	9	87	4	Schnurre, 1956
	182	36	63	2	Fischer, 1970
Germany	?	12	60	18	Fischer, 1970
Romania (middle Danube)	?	79	21	–	Fischer, 1970
Great Britain	141	38	51	11	Love (unpublished)

Fig. 40. Adult White-tailed Sea Eagle landing at eyrie with a rough dab (*Hippoglossoides platessoides*) for the eaglet, Greenland. (Photo: F. Wille)

It is during the winter months that White-tailed Sea Eagles, both adults and immatures, seem most willing to utilise carrion. Last century, 70% of winter observations which have come to my attention were of Scottish Sea Eagles taking mammal prey. Since these reports were gleaned from the literature, however, they may not be truly representative. Nonetheless, prey collections made in Norway (Willgohs, 1961) and in East Germany (Dornbusch, 1977) (Table 11) tended to confirm that mammal prey became more favoured in winter (bearing in mind too that carrion will tend to be underestimated among the food remains used in the analysis). The former study also revealed how birds (auks rather than Shag, Eider or gulls) became a more important constituent of the diet in winter when fish such as Cod, Lumpsucker and Catfish had moved to deeper water: being opportunists, the eagles could procure some fish by other means, such as carrion or as waste from man, otters, gulls, etc. Some of these eagles had been forced by winter snows and ice to the coast, where a spectrum of prey different to that of their breeding quarters was available to them – no freshwater fish, Grouse or small mammals as additional, if infrequent, variety. Helander (1981a) found that in southern Sweden birds became more important than fish in autumn and winter, while in Lappland reindeer carrion assumed some importance.

As one accumulates prey species lists, it becomes increasingly obvious just how catholic Sea Eagles are in their tastes. A cursory survey of the readily accessible literature reveals about 80 species of birds and over 30 species of fish, and a similar number of assorted mammal species have been recorded. Not surprisingly some species feature more than others, depending upon the hunting technique employed.

Fish

Most commonly the Sea Eagle hunts by gliding a few metres above the water, pausing to hover momentarily, before descending to strike. At the last moment the feet are brought forward to snatch the prey, with barely a splash (Figs. 41 and 42). Glutz von Blotzheim *et al.* (1971) claim that when searching from a height of 200 m or more, the sudden swift stoop terminates in a full immersion similar to that of the Osprey, but Willgohs (1961) asserts that plunge-diving is a rare occurrence. It is not unusual for a hungry Sea Eagle to operate from a perch overlooking water – a convenient crag, hillock or tree. On the northern coasts of Norway high tussocks on low, grassy islets are used. Repeated defaecations by the eagles encourage grass growth so that these traditional perches – termed 'ørnetuer' – grow bigger each year. A Sea Eagle may sit on the shore or even wade into the shallows seeking fish as would a Heron (*Ardea cinerea*). Willgohs (1961) also recorded an instance of two eagles sitting on open water, apparently attracted by a shoal of small fish: both birds took off without difficulty.

If the Sea Eagle is unable to plunge-dive as effectively as the Osprey, it procures deep-water fish by other means, not least by blatant piracy,

Food habits

Table 11. *Seasonal variation in diet of White-tailed Sea Eagles*

Country	No. of items	% fish	% birds	% mammals	Source
Norway					
winter	65	17	77	6	Willgohs, 1961
summer	99	37	60	2	Willgohs, 1961
East Germany					
winter	?	6	79	15	Dornbusch, 1977
summer	?	56	37	7	Dornbusch, 1977

Fig. 41. About to strike.

Fig. 42. Adult lifting a fish.

notably from the Osprey itself. It readily takes fish landed by Greater Black-backed Gulls (*Larus marinus*) and Herring Gulls (*L. argentatus*), especially Lumpsuckers from which the gulls characteristically had eaten only the eyes and the guts. Glutz von Blotzheim *et al.*, (1971) also mentioned how food may be stolen from Snowy Owls (*Nyctea scandiaca*), Red Kites (*Milvus milvus*), Black Kites (*M. migrans*), Buzzard (*Buteo buteo*) and Peregrines. I have also watched both Ravens (*Corvus corax*) and Hooded Crows (*C. corone*), being deprived of food.

As long ago as 1774, Low (1813) noted Orkney Sea Eagles lifting fish from otters (*Lutra lutra*). Willgohs (1961) was told that otters contributed much to the diet of Sea Eagles. Norwegian hunters often learnt of the whereabouts of otters by observing the behaviour of eagles, one describing how a patient eagle would glide down from its look-out nearby whenever an otter came ashore with a fish. Whether the mammal acquiesces or whether the eagle profits only after the otter has taken its share (usually having eaten the head, guts and liver) is debatable. In 1981 on a Scottish loch, Roy Dennis (personal communication) watched a first-year Sea Eagle (released on Rum 6 months previously) swooping very low over an otter and her two cubs. They dived and, once the eagle had perched some distance away, the female otter caught an eel which she was able to devour undisturbed on the surface of the water.

Nor is the Sea Eagle averse to scrounging fish from boats, around harbours or fish markets. As many as eight eagles have been seen to join the gulls following boats while the crew guts the catch. Willgohs (1961) mentioned an unusual instance of a Sea Eagle found dead in a trout net,. from which it had presumably attempted to steal. In 1883 Seebohm wrote how the 'species often made considerable havoc in carp ponds', while in the Sava valley of Yugoslavia, Suetens & von Groenendel (1968) observed a pair which fed their 8–9-week-old brood almost entirely on decapitated carp.

Thus, by various means a wide variety of species can be taken by Sea Eagles. Willoghs (1963) listed 328 fish of 24 species taken in Norway; nearly 50% of the sample comprised only two species – Lumpsucker (*Cyclopterus lumpus*) (24%) and Catfish (*Anarhichus lupus*) (17%): Cod (*Gadus morhua*) and other Gadids constituted a further 27% and Sea Perch (*Sebastes marinus*) another 10%. In a shorter list of 91 specimens of 9 species from Greenland, Kampp & Wille (1979) found 56% to be made up of Gadids and 33% of Char (*Salvelinus alpinus*); among the remainder were 5 Scorpion Fish (*Cottus scorpius*) and single specimens of *Cyclopterus* and *Anarhichus*. Seventeen Gadids (52%) featured among 33 fish listed by Flerov (1970) from the White Sea area, with 7 each of *Anarhichus* and flatfish (*Pleuronectes*) (each 21%) and two Herring (*Clupea harengus*). Catfish and Lumpsuckers appear also to be common prey in Iceland (Lewis, 1938). A dozen Cod were found in one eyrie in the Isle of Skye while at another a local shepherd recalled that Grey Gurnard (*Eutrigia gurnardus*) were frequent (Harvie-Brown & Macpherson, 1904). Skate (*Raia batis*) and Halibut (*Hippoglossus*) have been recorded in Shetland eyries in the past, Saxby (1874) going on to describe how the Erne was supposed to deal with a large specimen: 'He raises one of his wings, which serves as a sail and if favoured by the wind, in that attitude he drifts towards the land. The moment he touches the shore, he begins to eat out and disengage his claws, but if discovered before this can be affected, he falls an easy prey to the first assailant.' This tale may not be so fanciful as it sounds for Wille (1979) has photographed a Greenland Sea Eagle 'rowing' ashore with its wings whilst still clutching a fish which was too large for it to lift from the water (Fig. 43). Indeed it has been claimed that eagles can drown in this manner and the number of such cases documented (Willgohs found 35 in Norway alone) makes it difficult to discount them as merely fiction. Most involved Halibut or Salmon (*Salmo salar*). In one or two accounts the witness claimed to have heard a loud crack – the bird's wings breaking as it was dragged beneath the surface. Buckley & Harvie-Brown (1891) were told of a large Halibut that was found with an eagle's feet still embedded in its back, the bird having rotted off! Ospreys may similarly succumb, although it should be mentioned that a photograph was published in the journal *British Birds* which purported to show the skeleton of an Osprey with talons firmly embedded in the body of a Carp: G.S. Cowles (1969) of the British Museum later demonstrated this to have been a hoax; not only had the bird been attached artificially after the fish had died but the skeleton actually belonged to a Buzzard!

Whether all these accounts and observations are accepted or not they do raise the problem of how heavy a fish a White-tailed Sea Eagle can actually carry. Willgohs (1961) quoted Welle-Strande who claimed that an eagle had lifted a Halibut weighing 15 kg, but described an instance of another eagle which dropped a Cod weighing only 8 kg. It would seem, however, that most fish caught weigh from 0.5 to 3 kg and that the smallest are Herring or Scorpion Fish, 15–20 cm long. Small fish feature rarely in the diet.

In freshwater, a completely different range of species is available to the Sea Eagle and by far the most commonly caught fish is the Pike (*Esox lucius*), a species much given to swimming immediately beneath the surface. Flerov (1970) quoted specimens weighing from 2 to 5 kg, with lengths of 60–90 cm, as being most frequently taken. Fischer (1970) listed 60 Pike taken by Ernes on the Muritz river in East Germany, together with three Bream (*Abramis*

Fig. 43. Adult swimming ashore with heavy prey. (From a photograph by F. Wille).

brama), a Perch (*Perca fluviatalis*) and an Eel (*Anguilla anguilla*). Schnurre (1956) identified 36 fish taken from eyries on the Darss Peninsula, also in East Germany, and which included eight Pike, 14 Eels, seven Cod (taken offshore) and a Perch. Carp, Trout, Shad (*Alosa alosa*) and Roach (*Rutillis rutilis*) have also featured as prey. In the Quarken Straits of Finland, Pike constituted 72% of the fish diet, with Perch, Roach, Bream and Orfe (*Leuciscus idus*) making up the remainder (Stjernberg, 1981). Ussher & Warren (1900) described how Sea Eagles in Ireland would often watch over fords where Salmon leapt, although dead or spent fish were also taken.

Thus, although often preferring to catch only a few species, Sea Eagles will at times be tempted by many others. Willgohs (1961) aptly concluded that 'The eagle will probably take almost any fish species of convenient size where it can easily be obtained, and therefore the fish diet may vary considerably from place to place and from time to time'.

Birds

Similarly, particular bird species tend to be favoured, although some 80 species have been recorded, and again Sea Eagles can employ diverse techniques in their capture.

Meinertzhagen (1959) watched an eagle being violently mobbed by gulls until, with a sudden and elegant twist, it succeeded in grasping one reckless individual which it then plucked and ate on a nearby post. A Goshawk has been seen to succumb to an eagle in like manner (Willgohs, 1961) but on such occasions the eagle is perhaps displaying more good luck than judgement. Rudebeck (1951) observed a Heron (*Ardea cinerea*) being caught in flight and a Mallard (*Anas platyrhynchos*) being struck in mid-air: aerial attacks on geese were unsuccessful. An element of surprise, together with a tactful use of cover or bright sunlight, may enhance an eagle's chance of success. Willgohs (1961) noted an eagle fly low over the water to snatch an unsuspecting Heron on the shore, and I have seen an eagle on Rum employ a similar technique with the added advantage of bright sunlight behind, while attempting to surprise a flock of gulls on the shore. Moulting geese can be caught with relative ease. Like most predators the Sea Eagle is also ready to capitalise upon any debilitated animal. Ducks injured by hunters are easy prey while both Mute (*Cygnus olor*) and Whooper Swans (*C. cygnus*) frozen into the ice in lakes have been taken (Fischer, 1970). Rudebeck (1951) saw an injured Curlew (*Numenius arquata*) attacked (although he did not witness the outcome); an injured Bean Goose (*Anser fabilis*) successfully defended itself with its wing. Rudebeck witnessed only three successful hunts, and one victim of these was a Mallard unable to fly. He claimed to have observed 60 'hunts' altogether, so only 5% had proved successful. This seems exceedingly low and may result from including many low-intensity stoops which even the eagle itself may not expect to prove fruitful. The deliberate selection of debilitated prey is interesting in itself, however, and is better demonstrated by Rudebeck's observations on Sparrowhawks (*Accipiter nisus*), Merlin (*Falco columbarius*) and Peregrine where he was able to recognise injured or otherwise abnormal individuals amongst about 20% of the prey taken. By being selective to increase its chances of success, a predator is performing a useful 'sanitary' function on the prey populations.

One of the young Sea Eagles released on Fair Isle in 1968 was seen to catch a Fulmar (*Fulmarus glacialis*) in flight but released it almost immediately (Dennis, 1969). Within a few weeks, however, it was achieving regular success. Several of the Rum eagles also learnt to catch both gulls and Fulmars in flight, the latter seemingly possessed of a suicidal curiosity! Fulmar chicks may be approached as they sit on the nest ledge but the eagle renders itself liable to being spat upon. This foul oily liquid proved the ultimate undoing of one of the Fair Isle eagles, as we shall see in a later chapter. Both in Norway and on Rum, Sea Eagles will visit seabird colonies to lift young Kittiwakes (*Rissa tridactyla*) from nests. Uttendorfer (1939) reported that heronries were similarly raided, also the nests of Rooks (*Corvus frugilegus*), Black-headed Gulls (*Larus ridibundus*) and even nestlings of Osprey, hawks (*Accipiter* spp.) and Black Kites. Young birds are, of course, especially vulnerable and 86% of the Black-headed Gulls killed on the East German lakes were juveniles (Fischer, 1970). Wolley (1902) noted 'heaps of young Herring Gull remains' at an eyrie on Dunnet Head in Scotland. Soot-Ryen (1941) watched Greater Black-backed Gulls (*Larus marinus*) in Norway snatching young seabirds on their maiden flight to the sea and in turn being pursued by piratical Sea Eagles.

Exceptionally, Sea Eagles may take eggs which, according to Glutz von

Blotzheim *et al.* (1971), may be carried in the beak. In Norway it is suspected that the eggs of Kittiwakes, Eider (*Somateria mollissima*), Shag (*Phalacrocorax aristotelis*) and Gulls (*Larus* spp.) have been eaten (Willgohs, 1961). Both Grenquist (1952) and Flerov (1970) recorded incubating Eider being killed and two such instances have been noted recently on Rum (Love, 1980b). Eider ducks flushed from a nest fly ponderously and perhaps weakly so that they are easy prey (Glutz von Blotzheim *et al.*, 1971).

Typically, however, a Sea Eagle hunts low over a flock of birds on the water, repeatedly swooping to attack one luckless individual whose only escape is to dive again and again until, exhausted, it emerges for its final breath (Fig. 44). Flerov (1970) watched an eagle make seven such attacks on one Eider, and 12 on another. (In neither instance could the eagle lift its prey from the water but instead dragged it to the shore.) Glutz von Blotzheim *et al.*, (1971) recorded up to 65 attacks being made in periods of 35 to 45 minutes. Not surprisingly, such efforts can exhaust the eagle too, and one immature had to give up after 15 and 28 attempts at Dabchick (*Podiceps ruficollis*).

After an unsuccessful stoop at a diving bird, the eagle circles low to keep close to the bird when it next emerges. It is easier to follow the prey beneath the surface in shallower water, and Willgohs (1961) suggested that male Eiders more frequently fall victim because they are more conspicuous under water. A pair of eagles sometimes proves to be a more effective team, one being near the surface when the victim emerges, the other remaining at a suitable height to watch the movements of the prey under water and to assume position for the next attack. The Swedish artist Bruno Liljefors (1860–1939) vividly painted two Sea Eagles attacking a Red-throated Diver (*Gavia stellata*), a scene strikingly similar to one painted by J.G. Millais, who claims to have witnessed two eagles tiring out a Great Northern Diver (*Gavia immer*) near Lofoten. In his book *A reed shaken by the wind*, Gavin Maxwell related how he watched several thousand Coot (*Fulica atra*) bunching together in panic on a Euphrates marsh, attempting to evade repeated attacks from five Sea Eagles.

Carrying a Coot, an eagle is able to tuck up its legs, but with larger prey its legs hang down. Both feet may be required to carry larger prey, and only one for small prey. Glutz von Blotzheim *et al.*, (1971) note several instances where one Eider chick was carried in each foot.

Repeated harrying of birds on the water is most successful with diving birds, and consequently such species figure prominently in the diet. Eider constituted 30% of avian prey items in Norway (Willgohs, 1963) and around the White Sea in the USSR (Flerov, 1970). During the winter months in Finland, Sea Eagles were attracted by the huge concentrations of Long-

Food habits

Fig. 44. Striking at a surfacing Eider drake.

tailed Duck (*Clangula hyemalis*) and Goldeneye (*Bucephala clangula*) (Bergman, 1961). Other species occurring in the Norwegian sample were Shags (10%), Guillemot (*Uria aalge*) and other auks (28%), with gulls forming another 13%. Eider and Glaucous Gulls (*Larus hyperboreus*) were common prey in Iceland with some Fulmars, Ptarmigan (*Lagopus mutus*) and an occasional Merganser (*Mergus serrator*) (Ingolfsson, 1961). In Greenland, Eider were again favoured (20% of the avian prey) and to a lesser extent Long-tailed Ducks (6%); auks would seem to have been uncommon in the area so gulls (*Larus glaucoides* and *L. canus*, together amounting to 26% of the prey items), Ptarmigan (20%), Ravens (13%) and Red-throated Divers (8%) all featured instead (Kampp & Wille, 1979). Landbirds play a minor role in the diet of Norwegian Sea Eagles but in the White Sea a variety of grouse and waders, especially Snipe (*Gallinago gallinago*), were taken.

On freshwater lakes, diving birds again feature prominently. In the Quarken Straits area of Finland, Red-breasted Mergansers (*Mergus serrator*),

Goosander (*M. merganser*), Great Crested Grebe (*Podiceps cristatus*), Black-headed Gull, Mallard, Goldeneye, Velvet Scoter (*Melanitta fusca*) and *Aythya* spp. were taken (Stjernberg, 1981). In East Germany, Coot were important – 16% of avian prey listed by Fischer (1970) and 70% of that listed by Schnurre (1956). Mallard made up 19% and 16% in these respective samples. Dabbling ducks are less prone to capture by Sea Eagles because their alarm reaction is to take flight immediately. Individuals can sometimes be snatched from dense, panic-stricken flocks, but Fischer (1970) claims that only isolated birds are attacked. Certainly disadvantaged birds are easy prey, especially those injured or in moult or 'pelleted' birds which have escaped from wildfowlers. Fischer (1970) found that gosling Grey Lags (*Anser anser*) (23%) and juvenile Black-headed Gulls (25%) were taken, together with diving species such as Coot, Grebe, etc.

Willgohs (1961) was told of an eagle which pushed into a dense flock of Starlings (*Sturnus vulgaris*) to emerge with one in its talons. This and the Thrush (*Turdus* sp.) mentioned by Flerov (1970) are the smallest recorded prey taken by Sea Eagles, but seem to be taken infrequently. Such small items tend to be under-recorded since they leave few conspicuous remains, but they reward an eagle little food value anyway for the effort involved in catching them. Species such as Sandpipers (*Calidris*) (Witherby *et al.*, 1943) and Turnstone (*Arenaria interpres*) (Willgohs, 1961) are killed rarely but most prey species range from 0.5 to 2.5 kg in weight. The largest are swans up to 10 kg and Great Bustard (*Otis tarda*), about 15 kg.

A similar picture emerges when one scans old Scottish records. They include gulls (mainly Herring, Greater Black-backs and Kittiwakes), Guillemots and Puffins (*Fratercula arctica*). Wolley (1902) was told of 'thousands of cormorants' being found in one eyrie (possibly properly referring to shags?). As in Norway, gamebirds were rarely taken and Booth (1881–87) admitted that Sea Eagles were less destructive in this respect than Golden Eagles. Domestic fowl were easy prey and Dresser (1871–81) lamented how in the Highlands frequent onslaughts on the farmyard by Sea Eagles apparently caused much anxiety among the local inhabitants.

Other Prey

Before going on to consider mammalian prey and in particular domestic stock, brief mention should be made of a few other items which have appeared in the diet of Sea Eagles. Frogs and small reptiles have been recorded, including an Adder (*Vipera beris*), although the latter was immediately rejected (Willgohs, 1961). Captive Sea Eagles in Israel (Fig. 57) were seen to catch and eat small Swamp Turtles (*Clemmys caspica*) which frequented the ponds in the aviary (Edna Gorney personal communication).

Uttendorfer (1952) noted a Sea Eagle seeking molluscs and snails on the banks of the Rhine, while according to Glutz von Blotzheim et al., (1971) another was seen robbing Crows of freshwater mussels (*Anodonta*) they were dropping; any unopened were dropped by the Eagle in a similar fashion but from a height of only a metre or so above the ground. Some reports suggest that cuttlefish, crabs, lobsters, starfish and sea urchins have been taken. One eagle was said to have been gorged on jellyfish (*Aurelia* or *Cyanea* sp.) but Willgohs (1961) suggested that this may have been a diseased or injured individual. Miscellaneous shells of marine mussels (*Mytilus*) and snails (*Gibbula*, *Buccinium*, *Littorina*, etc.), occasionally found in nests, at roosts, or in pellets most likely originate from the stomachs of Eiders.

One further item – human babies – said to have been taken by eagles, instantly arouses horror, but merits careful and critical assessment. Tales from the past of eagles snatching infants would appear to be legion; few are credible. With almost monotonous regularity the unfortunate child was rescued unharmed through the heroic efforts of its distraught mother. One determined Faroese woman was said to have scaled the inaccessible crags of the island of Tindholmur (Fig. 45) to reach her baby – a feat never before accomplished even by the most daring of local cragsmen (Williamson, 1970). Alas the baby was dead, its eyes apparently pecked out. If in truth the event did take place, this grisly detail would seem to exonerate the Sea Eagle; the child must already have been dead and the victim of a crow attack before it was found by the eagle.

Careful attention to detail can often make a tale sound more plausible and was perhaps a deliberate ploy, additive over several generations of narration. Tulloch (1978) was told at considerable length of an infant from Unst in Shetland, having been carried by a Sea Eagle to its eyrie on the nearby island of Fetlar. Eventually a local lad descended the cliff to rescue the baby girl. Many years later when she grew up she became the bride of her bold rescuer, thus fulfilling the prophecy which had been uttered by an old man who witnessed the scene. This happy event, however, apparently occurred nearly 200 years ago. This is a feature all too common to 'eagle and child' stories. None have taken place within living memory with one notable (although not necessarily truthful) exception.

Dr Johan Willgohs (1961) was able to interview a Norwegian woman who claimed, as a child in 1936, to have been carried off by a Sea Eagle. Again the detail sounds convincing except that at the time she had been $3\frac{1}{2}$ years of age. Whether an eagle would ever have succeeded in carrying the child's weight of 19 kg (42 lb) to a height of 200 m and a distance of 3 km, to the spot where its eyrie was located, must be severely open to question. It is

Fig. 45. The pinnacle of Tindholmur, with the island of Mykines in the background, Faroe Islands.

true that Sea Eagles may lift weights of up to 11 kg, and once a load of 15 kg (but, significantly, another eagle had to drop a dog of this weight almost immediately (Willgohs, 1961)). Only the largest eagles could ever contemplate such feats and even then would require all help from favourable topography and winds.

Thus, none of these claims would appear to stand up to critical scrutiny and all such eagle and child stories should remain firmly within the realms of folklore. Some have been preserved for posterity as attractive, though sometimes heavily stylised, inn signs (Fig. 46). With tongue firmly in cheek, Robert Gray (1871) observed that 'it is perhaps unnecessary to congratulate ourselves that through the diligence of keepers and collectors, we are spared the infliction of seeing a modern perambulator relieved of its occupant'. Since the advent of firearms, eagles have developed a healthier respect for man and his habits and rarely are modern children left out in such vulnerable and exposed situations. Indeed one eminent professor cynically remarked that if an infant should be borne away it is the mother who is to blame and not the eagle.

Food habits

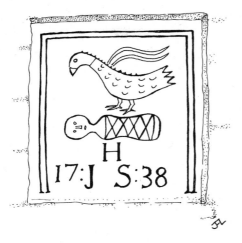

Fig. 46. 'Eagle and child' sign in the wall of an inn near Allgreave, Cheshire.

Mammals

In Arctic regions during years of population explosions of voles or lemmings, these may be caught by Sea Eagles; half of the mammal prey species noted by Flerov (1970) in the White Sea area were voles. Willgohs (1961) recorded an occasional Water Vole (*Arvicola terrestris*) in pellets collected from the outer skerries of Norway. One of his correspondents witnessed an eagle hover momentarily to pounce on a vole on the ground. Willgohs quoted Marmots (*Marmota*), Susliks (*Citellus*) and Mole Rats (*Spalax*) as prey in the Russian steppes. Rats (*Rattus norvegicus*), Field Mouse (*Apodemus*) and other mice, hamsters, Moles (*Talpa europaea*), Hedgehogs (*Erinaceus europeus*) and Squirrels (*Scurius vulgaris*) have all been found at eyries (Willgohs, 1961). Berg (1923) once saw a Sea Eagle near its eyrie catch a Red Squirrel in a tree. Muskrats (*Ondatra zibethica*) and Mink (*Mustela vison*) featured to a minor extent in the diet of White-tailed Sea Eagles in the Quarken area of Finland (Stjernberg, 1981). On the Ili delta of Kazakhstan (USSR), Muskrats from local farms were taken in the spring and autumn (when they constituted 30% and 43%, respectively, of all items found at Sea Eagle eyries): for some reason they were less favoured or unavailable in summer (14% of items) when voles were caught instead, although to a lesser extent (Glutz von Blotzheim *et al.*, (1971).

Probably the most common live mammal prey taken by Sea Eagles are Rabbits (*Oryctolagus cunniculus*) and hares. Arctic Hares (*Lepus timidus*) seem to be especially vulnerable in their white winter coat when there is little snow or where cover is sparse, as on offshore islands in northern Norway (Willgohs, 1961); in Greenland, the equivalent, *Alopex lagopus*, appears as prey of Sea Eagles (Kampp & Wille, 1979). It is not unknown for two or three birds to operate together to catch hares (Willgohs, 1961). In Britain, Brown Hares (*Lepus europaeus*) and Rabbits were common winter prey (Macgillivray, 1886; Gray 1871; Harvie-Brown & Buckley, 1892; Macpherson, 1892; Forrest, 1907), while both Common and Arctic hares were occasionally taken during the breeding season (Blackburn, 1895; Ussher & Warren, 1900; Forrest, 1907; Jourdain, 1912).

Macgillivray (1886) asserted that Sea Eagles were 'especially fond of dogs'. Willgohs (1961) recorded how one unfortunate hound of 11 kg was carried off but another of 15 kg was too heavy and had to be dropped. Cats may at times also fall victim (Pennant, 1774; Bewick, 1821), together with a variety of small carnivores such as martens. It would seem that Otters (*Lutra lutra*) are rarely attacked, although Willgohs (1961) mentions an instance in Norway and another in Sweden, while Uttendörfer (1939) mentioned a bitch otter being killed just as it gave birth to its two young. Dick (1916) related an old account of a fight between an eagle and an otter on a loch in southern Scotland. He did not think they were contesting a fish, a second eagle appearing on the scene and managing to sink its talons into the otter before it finally escaped. A live Jackal (*Canis aureus*) which had been caught in a trap was taken by a Sea Eagle (Bannerman, 1956) while a Silver Fox was killed after having escaped from a fur farm in Norway (Willgohs, 1961). Two foxes are said to have been lifted by eagles during organised 'fox drives' in Norway – doubtless much to the annoyance of the hunters (Willgohs, 1961). There are also the inevitable tales of furious battles between eagle and fox, and in Norway a fox skeleton has been found with the bones of an eagle firmly attached (Willgohs, 1961).

An eagle was once seen to lay hold of a sleeping seal which instantly dived taking the bird with it; shortly afterwards the eagle reappeared with its wings broken and on the point of death (Bannerman, 1956). Attacks on young seals are not uncommon, Seton Gordon (in Bannerman, 1956) pointing out how an eagle might easily mistake a sleeping seal for a dead one. Willgohs (1961) recorded a Sea Eagle drowning in an attempt to catch a live Porpoise (*Phocoena phocaena*)! In the main, however, seals and cetaceans feature in the eagle's diet only as carrion. The same may be said for most ruminants, although calves of Roe Deer (*Capreolus capreolus*) or Reindeer (*Rangifer tarandus*) can be taken alive. There are only one or two records of full-grown deer being killed. One panic-stricken Roe was seen to collapse exhausted with an eagle clinging firmly to its back (Willgohs, 1961) but another more fortunate beast on the Isle of Harris successfully evaded capture (Macgillivray, 1886).

Predation upon domestic stock

There was at one time a belief in Shetland that eagles which were tempted to raid the farmyard could be prevailed upon to drop the victim by a charm; the bystander was supposed to cast some knots in a length of string and utter a simple spell. It was prudently added that the successful outcome might only ensue some distance away; so that the charmer could evade embarrassment no doubt! No less eminent a personage than the local minister from the town of Scalloway was said to have witnessed one such successful spell; in the absence of string the aspiring charmer was seen to make do with his garter (Ritchie, 1920).

Despite the dubious efficacy of this antidote, the periodic appearance of domestic stock on the diet sheet of the Sea Eagle has incurred the bird much enmity from man. Piglets have on rare occasions been lifted (Macgillivray, 1886) although C.M. McVean, according to Gray (1871), claimed that his tame eagle was afraid of pigs. Goats (*Capra hircus*) have also been attacked and indeed the carcass of almost any animal – even cows (Fergusson, 1885; Willgohs, 1961) – may be attractive to a hungry eagle. At certain times of year there may be an abundance of carrion available. In Argyll, on the west coast of Scotland, for instance, some 6 or 7% of Blackface ewes (*Ovis aries*) die each year (Houston, 1977); not all are buried as is demanded by law. Thus it is not surprising that along a 2-mile transect in Lewis, Lockie & Stephen (1959) encountered no fewer than 28 dead sheep. Overstocking in the past may have contributed to a gradual deterioration in grazing quality, but even a century ago a similar situation prevailed in the Highlands, as Booth (1871–81) attested: 'After a protracted winter and a dry cold spring the herbage is scarce, so many dead carcasses may be found in all directions, scattered over the moors'.

Lambs too are especially vulnerable, and with the availability of dead or dying ones it is understandable that an eagle may occasionally be tempted to lift one still alive. When a poor shepherd witnesses his livelihood disappearing thus, one can appreciate his animosity. It may be only a few 'rogue' eagles which bring upon their innocent peers an ill reputation, but it can rapidly generate a determined campaign of destruction. Macpherson (1892) expressed a typical sentiment – from the Lake District, but one that could be found anywhere that eagles were to be encountered: '. . . these birds were so destructive to the interests of the shepherds, that their extermination became absolutely necessary'.

Clearly, anyone advocating the reintroduction of the White-tailed Sea Eagle to our shores must assess critically such claims. One's stance will differ according to one's interests, whether one is a sheep farmer or an ornithologist; as an ornithologist I shall attempt to approach the con-

troversy with some degree of objectivity. Four considerations have to be borne in mind:
1 The importance of lambs in the diet of the Sea Eagle.
2 The number of lambs which were taken alive by the eagles.
3 The proportion of these which would have died anyway.
4 The significance of this eagle-induced mortality, relative to the total lamb deaths in the whole flock.

Now that the Sea Eagle has been exterminated from Britain we have to take a rather indirect approach to the problem; by studying the writings of people who had direct experience of eagles and lambs last century, by examining the situation abroad where Sea Eagles are still to be found, or by extrapolating from the habits of a similar species – the Golden Eagle – still common in this country. In a survey of the old ornithological, oological and sporting literature I have gleaned a total of 141 items which had been recorded at eyries of the Sea Eagle prior to its demise in Britain. Only 12 (9%) of these observations involved lambs. Seven of them were found at one nest in the Lake District (Macpherson, 1892); this serves to demonstrate how the problem may be of a local nature. Indeed several Sea Eagle pairs merited specific mention in the literature because they never touched lambs (Martin, 1716; Macaulay, 1764; Venables & Venables, 1955). When Lord Lilford (1885–1897) was travelling through Greece in 1860, he made to raid one Sea Eagle eyrie but 'the shepherds begged us not to kill them as they bred year after year and kept away other birds of prey which were destructive to their lambs'. With a knowing smile Lilford was disinclined to believe them but, to his credit, 'scrupulously attended to their request'. On a similar theme Lockie (1964) found lamb-killing by Golden Eagles to be a local problem only; it was prevalent in those areas of northwest Scotland where the eagles' preferred prey – grouse and hares – was scarce.

Returning to the old literature, there are a further 46 observations of Sea Eagles feeding at prey away from the nest, mostly in the winter months. Sheep and lambs feature in only six of them. Four involved eagles feeding at adult sheep which must certainly have been discovered as carrion. One lamb is recorded as being 'lifted' by a Sea Eagle, but it was immediately dropped again unharmed (Harvie-Brown & Macpherson, 1904). Another instance is detailed where 10 or 12 lambs were picked up one after the other but each dropped at once – again all unharmed. Eagles on one small island in Shetland were accused of killing seven lambs 'in a short time' (Evans & Buckley, 1899). Two further reports, however, I have dismissed as being exceedingly unlikely and mere fable. One pair of eagles in the Outer Hebrides 'came daily from Skye with a young lamb for each of their eaglets'; another pair were said to have undertaken the reverse journey, from Barra

Head to Tiree, – a distance of 40 miles! Both tales are included in Fergusson's account of the Gaelic names and folklore of birds, and the latter story in particular, with all its entertaining embroidery, seems entirely apocryphal. In brief, the lamb survived the ordeal and being of a distinctive breed was recognised many years later by, of course, his former owner on a fortuitous visit to Tiree; he was gratified to learn that his lost ewe had had a long and very productive life.

One discovers in the literature many other vague accusations of lamb-killing by eagles but in this critical and scientific assessment we must lay aside such hearsay. Some may indeed be based upon fact and to be fair we must remember that the shepherds who were the most likely to witness such depredations on their flock were least likely to record their experiences on paper, but it can be all too easy to overstate the problem. Certainly, studies in Norway and elsewhere in recent times would bear this out. Willgohs (1961) gathered 36 instances of lambs or goat kids being taken by eagles, but only 12 could be said to have been killed by eagles. In a further five instances it was positively known that the animals had been taken dead. Many of the reports emanated from the southern counties of Norway, with relatively few further north where eagles were much more common, and where it was frequently intimated to Willgohs in the course of his researches that the eagles were harmless. In a list of over 1400 prey items noted from Norwegian eyries, only 60 (4%) were sheep and lambs. While it is true to say that carrion is difficult to identify with certainty amongst such prey remains, it is also clear that sheep and lambs, whether live or dead, are of very minor importance, numerically, in the diet of the Sea Eagles.

In terms of bulk, however, such beasts assume greater importance since they are so much larger than many other prey items. In Western Ross, where Golden Eagles were taking an unusual number of lambs, Lockie (1964) calculated that they comprised some 30% by weight of the eagles' intake. He found remains of 22 lambs in one eyrie, over five consecutive seasons, and on 10 of the carcasses he was able to carry out a useful *post mortem*; seven had been alive when lifted by the eagles, as evidenced by the bruising and bleeding under the skin where it had been pierced by the talons. This can be viewed in better perspective when one considers that there were 1000 ewes breeding in the territory of that eagle pair. Weir (1973) estimated that another pair of 'rogue' eagles in the Highlands had taken some 20 lambs during the period that their eaglets were in the nest: this made up some 75–85% by weight of the prey found at their eyrie, but even in this instance, it represented only some 2% of lambs born within the hunting range of the eagle pair. Only two of four carcasses examined could be claimed to have been killed by the eagles. It is however also important to assess how many of these lambs were in fact healthy and would otherwise have survived. Houston (1977) investigated lamb mortality in Argyllshire, where some 15–20% of lambs die, mostly within their first week of life. He examined the carcasses of 254 of them to determine the causes of death: 27% had been still-born, 5% had died almost immediately after birth and before they had walked, 9% were diseased, and another 5% were killed by accident (having been sat upon by the ewe, strangled in fences, hit by cars or killed by dogs). Amongst the remainder for which no cause of death was apparent, nearly all of them (45% of the entire sample of 254) were found to have exhausted fat levels – they had effectively starved; 75% of them had never even sucked. A few of the others had 'reduced' fat levels, and only 7% seemed healthy. This sad state of affairs would appear to have been by no means a recent one in the Highlands. As long ago as 1871 Gray wrote that '. . . it cannot be denied that the Sea Eagle oftener feasts upon carrion than upon living animals, and that in the most of cases where lambs are actually lifted the offence is to a great extent mitigated by the fact of the severe spring weather having previously crippled these poor creatures beyond hope of recovery . . .'.

It would appear that Houston's findings are not at all atypical in sheep husbandry. In Australia, for instance, Rowley (1970) found that 33–67% of dead lambs (excluding those still-born) were starving prior to their demise. He established that exposure can be a significant factor in the cause of death. Mortality amongst lambs kept in exposed conditions was about 35%, but once shelter was provided it could be reduced to only 4%. It is well known that the age and health of a ewe affect her lamb. She may be unable to produce enough milk to feed it or the lamb may be small or weak and therefore more susceptible to exposure. Also, Alexander *et al.*, (1967) discovered that better-fed ewes are more aggressive in the defence of their offspring from predators. Regular feeding of ewes and the bringing about of lambing on richer pastures were found to reduce mortality of Blackface lambs from 14 to 5% (Gunn & Robertson, 1963). 'Flushed' ewes, given extra feeding in weeks prior to their being put to the ram, tended to have a higher incidence of twinning. Although twin lambs may be smaller (and more susceptible to exposure) this increased breeding output means that one of the twins can easily be cross-fostered on to another ewe which has lost her lamb. Such improvements in sheep husbandry may not diminish the frequency of still-births, however. Indeed, difficult births may be more likely to occur if the lamb is a large one so that in addition to the possibility of the lamb dying the ewe herself may be put at risk.

Undoubtedly the application of such refined husbandry techniques (and in this must be included careful maintenance and improvement of hill pasture, together with rigorous culling, by the shepherd himself, of animals showing poor performance) will help increase lamb survival. It seems to me that benefits accruing from such efforts far outweigh any brought about by time-consuming, and usually illegal, predator control.

Wagner (1972) demonstrated that extensive poisoning of coyotes – a species considered by many American farmers to be a significant menace to their sheep – made no difference whatsoever to the annual losses of stock. This is in accordance with an idea, proposed by Errington (1946), that 'a great deal of predation is without truly depressive influence, in the sense that victims of one agency simply miss becoming victims of another . . .'.

With losses from other causes being so high in some parts of the Highlands, the abundance of carcasses would preclude the need for many lambs to be killed by predators. Lockie (1964) noted that lambs featured more in the diet of some Golden Eagles in years of poor lamb survival; in good lambing years the incidence of lamb remains found at eyries was halved, from 46% to only 23%.

The late Dr J.W. Campbell once commented: 'One is left with the impression that the threat normally to sheep under average conditions, was no greater from the Sea Eagle than it is today from the Golden'. I would prefer to claim that it is much less, because of the Sea Eagle's decided preference to carrion. Dixon (1900) astutely remarked that of the two species the White-tailed Sea Eagle 'is a regular scavenger of the shore . . . Healthy vigorous birds or animals are seldom attacked by this eagle: it confines its attentions to the weakly and the wounded creatures that cannot move fast or offer any serious resistance.' On the question of livestock, Macgillivray (1886) was more specific: ', , , at seasons of mortality among sheep, as in the end of autumn, when braxy commits its ravages, or in the end of spring, when severe weather often causes the death of young lambs, they [Sea Eagles] are not uncommonly seen hovering about'.

In summary, the conclusions of Gray (1871) seem appropriate: 'The chief food of this eagle appears to be stranded fish, procured in the vicinity of its maritime haunts, dead sheep found on the moors, and occasionally a salmon left by some scared otter – a selection more in keeping with the innocent life of a vulture than the plundering habits usually ascribed to eagles'. To this can be added an ability to catch fish on its own account, the Sea Eagle tending to snatch them near the surface; thus Lumpsuckers are frequent prey in coastal waters and Pike in freshwater. Although gulls, geese and Mallard may fall prey, diving birds such as Eider, auks and Shag are particularly vulnerable in coastal situations, with birds such as Coot on freshwater lakes. Mammals are rarely an important food but are sometimes taken as carrion. Sheep and lambs have generated more emotion among shepherds than they ever provide actual nourishment for eagles. The evidence would indicate that extermination of the Sea Eagle because of its depredations upon domestic stock was totally unwarranted – as we shall see in the following chapter, a deliberate and reprehensible act of vandalism.

6 Persecution and decline

> It is impossible to conceal the fact that if the present destruction of eagles continues we shall soon have to reckon this species amongst the extinct families of our 'feathered nobility'.
>
> R. Gray (1871)

In the Faroe Islands there once operated a very ancient tax system called 'Nevtollur' (Williamson, 1970). The custom demanded that every man with part ownership in a boat was obliged to pay yearly either one eagle, Raven or like bird of prey to an officer in the capital Torshavn. The eagle was especially valued, presumably because of its scarcity, and the man who presented the beak of an eagle was exempt for life from paying further 'bill tax' (Feilden, 1872). Indeed this bounty scheme may have played no small part in the early extinction of the Erne in the Faroes. The pursuit of White-tailed Sea Eagles nesting on the impressive cliffs of the Faroe Islands would have posed little problem for a people whose economy was based upon harvesting seabirds and fishing. There was also the local belief that the claws of the Sea Eagle could cure jaundice, but unless this affliction was unusually rife in the archipelago any demand for this commodity is unlikely to have been a significant factor in the species' demise there!

Bounty schemes were operated in many European countries. During the latter half of the nineteenth century, the Norwegian authorities met some 90 000 claims. Gradually the number of claims diminished (illustrating perhaps the effectiveness of the system) until it was finally abolished in 1932. At this time it stood at 2 Kroner per bird, but the provincial authorities and hunting clubs continued to offer payment so the slaughter continued. Between 1943 and 1950, 490 White-tailed Sea Eagles were killed on the island of Vikna alone, by which time as much as 50 Kroner could be claimed for every eagle killed. Between 1959 and 1968 an average of 169 eagles were being killed annually: a maximum of 221 in 1961 (Willgohs, 1969).

A similar slaughter was being effected throughout Europe whether bounties could be claimed or not. In Germany for instance, during the eighteenth century it is said that over 624 000 birds of prey were destroyed in the Hannover/Bremen region alone. The Sea Eagle figured prominently in such 'vermin' lists and around 1860 about 400 were being killed annually throughout Germany. In Sweden too, around the same time, over 200 Sea Eagles were being destroyed each year; Bijleveld (1974) cited countless other examples from all over Europe and they make depressing reading. There can be little doubt about the immense pressure that the species suffered from persecution, let alone the increasing loss of its habitat. It is perhaps surprising that it has survived at all.

In Britain the pattern was very much the same. From Ice Age times the White-tailed Sea Eagle had been a feature of the British landscape. Claws belonging to the species have been found in Tornewton Cave in South Devon along with the bones of Shelduck, Stork, Brent Goose, Wigeon, Teal and Goosander – all birds which are likely to fall prey to the Sea Eagle. It is possible that the entrance to the cave may have been the site of an eyrie or at least a roost. The deposits are ascribed to the third or Ipswichian glaciation and the succeeding warm phase (Harrison, 1980). The ice sheets advanced again during the Devensian glaciation, but during this time there were several brief mild periods in one of which deposits were laid down at Walthamstow in Essex. Here the bones of Grey Lag Goose, Mallard and Tufted Duck have been found, together with the left tibio-tarsus of a White-tailed Sea Eagle (Harrison & Walker, 1977).

In post-glacial times, remains of the eagle become more frequent. A claw appears amongst mixed Pleistocene and recent material from Cathole Cave on the Gower Peninsula in south Wales (Harrison, in press). Eagle bones have been excavated from a midden at Jarlshof, in Shetland, and date from the late Bronze Age/early Iron Age (Fisher, 1966a). The bird in question was almost certainly a White-tailed Sea Eagle, since one nested late last century at Fitful Head nearby and Golden Eagles have never been known to have bred in Shetland. The Sea Eagle would seem to have occurred in Orkney at this time. Sea Eagle bones have recently been excavated from a Bronze Age burial mound at Isbister in South Ronaldsay (J. Hedges, personal communication). At least 14 individuals were found in the mound along with many human burials and it is thought that the Sea Eagle held some totemic status to prehistoric man in Orkney. This cult might have persisted into the Iron Age for there is a magnificent eagle carving on a Pictish stone from the Knowe of Burrian. Its impressive outline is undoubtedly an Erne, being typically vulturine in form with a powerful beak and the diagnostic unfeathered tarsi (Fig. 47). Several other Pictish symbol stones in the North of Scotland show eagles, often in company with a large salmon. One stone from the St. Vigean's Museum, Angus, depicts an eagle in the act of devouring a huge fish (Fig. 47). In Celtic mythology the eagle is often associated with the salmon – the fish of knowledge – or else the bird is described as being perched on a sacred oak tree. An eagle, often clutching a huge fish, was adopted in mediaeval gospels as the symbol of the evangelist St. John, and our Burrian eagle reappears (Fig. 47) incorporated in a design from an eighth-century decorated manuscript (MS 197) in the Corpus Christi collection, Cambridge (Henderson, 1967).

Further south we again encounter Sea Eagles during the Iron Age. Their bones have been recovered from excavations of the lake village at Glastonbury (Harrison, 1980), on the Somerset Levels and at Lagore, Co. Meath (Fisher, 1966a). In addition, Fisher (1966a) listed over 40 different other species which represent a classic wetland avifauna – Cormorant, Heron, Dalmation Pelican, ducks, geese, swans, Coot, crane and gulls.

The Anglo-Saxons occupied much of southern Britain around this time, and were so familiar with the Sea Eagle as to make frequent reference to it in their literature. In the seventh-century poem *Judith*, appears

> . . . the dewy-feathered eagle,
> hungry for food;
> dark-coated, horny-beaked
> it sang a song of war.

After the Battle of Brunanburh in 937 AD we are told how the victorious Anglo-Saxon warriors

> . . . left behind them
> the black-coated raven, horny beaked
> to enjoy the carrion.
> And the grey-coated eagle, white-tailed,
> to have his will of the corpses.

During the Dark Ages the landscape of Britain had been little modified by man; huge tracts of deciduous forest flourished on the heavy lowland soils and extensive marshland existed around the Essex coast, the Fens, the Broads and the Somerset Levels. Such habitats would have been as attractive to Sea Eagles as they still are in some parts of Europe – the Baltic coasts of Germany and Poland, the complex river system of the Danube,

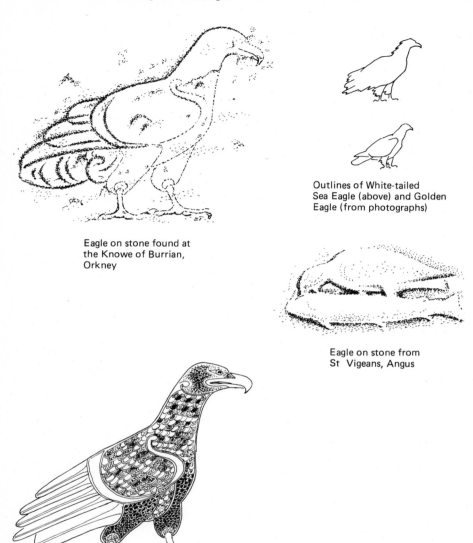

Eagle on stone found at the Knowe of Burrian, Orkney

Outlines of White-tailed Sea Eagle (above) and Golden Eagle (from photographs)

Eagle on stone from St Vigeans, Angus

Eagle in mediaeval manuscript (197) in Corpus Christi, Cambridge (8th century)

Fig. 47. Mediaeval representations of eagles.

etc., where relict populations of this bird yet survive. Waterfowl and fishes, such as Pike, abound in these lakes and marshes which are flanked by tall oak or beech stands to provide suitable eyrie sites.

Thus we can assume that the White-tailed Sea Eagle was once

Persecution and decline

widespread, if not common, in lowland Britain and Ireland. The Anglo-Saxons, however, had already begun to clear the lowland forests and to drain the fenland, processes which accelerated as the human population increased. On the newly created farmland 'noyfull fowls and vermin' were not tolerated and soon laws were introduced to encourage their destruction (Murton, 1971). Thus began the decline of the Sea Eagle in Britain. It is likely, however, that at this stage habitat loss was more significant than persecution, which was as yet on a comparatively minor scale. By the end of the eighteenth century, when sportsmen and naturalists were beginning to record the birdlife around them, only a handful of Sea Eagles remained in England – scattered pairs which sought a last refuge on the cliffs of the Isle of Wight (Yarrell, 1871), on Lundy, the Isle of Man and, probably, near Plymouth (Seebohm, 1883; D'Urban & Matthew, 1892; Bannerman, 1956).

Within a couple of decades, eagles in England were to be found only in the Lake District. Both Golden and Sea Eagles bred there, and from the literature it is sometimes difficult to differentiate which nested where, but the placenames survive – Eagle Crag, Erne's Crag, sometimes corrupted to Iron or Heron Crag (Ratcliffe, 1980), or even Wallow Crag (said to be from the Celtic *Iolaire* meaning eagle and pronounced 'yulir'). A Mediaeval document dated 1272 decreed that the small tenants of northern Cumberland 'must preserve the nests of Sparrowhawks and Eagles' (Mitchell & Robson, 1976), and Eagle Crag in Derwentwater was said to have been tenanted 'from time immemorial' (Macpherson, 1892). There the eagles had an eyrie 'far removed from gunshot, and undisturbed by men; for no adventurous fool ever dared assail their lofty habitation'. Premature and atypical sentiments, for the history of the eagles in Lakeland is rather one of relentless persecution. From 1713 to 1765, over 30 eagles were killed in Crosthwaite parish alone, with a bounty of sixpence being paid for a young bird and a shilling for an adult (Mitchell & Robson, 1976). Apparently the men of the parish kept a long stout rope at Borrowdale expressly for the purpose of climbing into the local eyries each year. At least one of these nests belonged to a pair of White-tailed Sea Eagles; about 1769 when a local farmer was lowered into the eyrie, the parent eagles circled overhead screaming anxiously, behaviour typical of this species, but not of the Golden Eagle. The nest was again robbed 3 or 4 years later, 'it being a common species of traffic in this country to supply the curious with young eagles' (Macpherson, 1892).

Seldom a year passed without the Lakeland eagles being shot, or their eggs or young being taken. One of the last nesting attempts by Sea Eagles took place on Wallow Crag near Haweswater in 1787 (Mitchell & Robson, 1976). A pair were seen in 1835 but they are not known to have nested. Golden Eagles ceased to breed there at about the same time and a stray seen in 1860 or thereabouts, according to Ernest Blezard, a local ornithologist, met a tragic if honourable fate. One Farmer Jenkins rammed 'a turrible gurt charge' of powder and lead down his muzzle-loader and lay down on his back to take aim at the eagle as it flew overhead. He fired and not surprisingly suffered a broken collar bone from the violent recoil; the eagle was only wounded and gave further account of itself before being finally despatched.

During the early part of the nineteenth century the Sea Eagle was still reasonably plentiful in the remote parts of Ireland and Scotland where the hand of man had little altered its coastal habitat. Here persecution was its sole worry, a practice which would seem to have had a long history and tradition in the area. There was an Act issued in Orkney and dated 1625 (Low, 1813)

> anent the slaying of the Earne . . . Whatsoever person or persones shall slay the earne or eagle, shall have of the bailzie [officer of justice] of the parochine [parish] qr it shall happen him to slay the earne or eagle, 8d for every rick [corn stack] within the parochine, except of the cottars who has not sheep; and 20s to ilk persone for ilk earne's nest it shall happen him to harrie . . .

Further north in Shetland it was stated that: 'Here are many ravenous Fowls, as Eagles, Ravens and Crows. In old times they so multiplied that the Fourde or Sheriff made an Act that whosoever at ye Head-Court brought in an Eagle's head, from each having sheep in that pasture he should have a merk . . .' (Venables & Venables, 1955).

By the eighteenth century, the *Old statistical account* for Orkney and Shetland tells us how the reward had risen to five shillings a head. This was ultimately reduced to three-and-sixpence before the system ceased altogether in 1835.

The bounties being offered by the authorities would appear to have been exceedingly generous; perhaps eagles were scarce or, more likely maybe, few people seemed motivated to claim them. Before the advent of firearms the methods available for destroying or capturing eagles would seem to have been either time-consuming or excessively hazardous. Few would be so lucky as the Shetlander who stumbled upon a Sea Eagle with its feet firmly trapped in the body of an enormous Halibut (Saxby, 1874), or the Faroese man who actually captured an eagle as it crouched innocently behind a rock sheltering in a snow storm – all he had to do was wring its neck (Feilden, 1872). Occupied eyries offered a convenient source of eagles for bounty claims except that they tended to be located halfway down sheer and well-nigh inaccessible precipices. Norwegian farmers prized accessible

Persecution and decline

eyries to supplement their incomes; eggs were sold to collectors and the eaglets sold to 'the curious' or for head-money. In classic farming tradition the adults would be spared so as to ensure an income for the following year (Willgohs, 1961). Every season the eaglets and one of the adults which had a nest by Lake Würm in Bavaria would be harvested to claim the premium (Bijleveld, 1974). It was only when a carcass of a royal deer was found in the eyrie that the peasants' regular income abruptly ceased.

Scottish crofters were no less enterprising but preferred simply to shoot whatever strayed within range of a nest, or else to place a trap in the nest beside the chick for the adults on their return (Blackburn, 1895). Sometimes a blazing barrel would be lowered to discourage the incubating bird, or else a clump of burning heather was lowered from above to set the nest and its contents alight (Dick, 1916; Macgillivray, 1886).

In both Norway and Scotland, before firearms came into cheap and common usage, an automatic trap was employed to catch eagles. It comprised simply a narrow trench with its walls partially built up with stones. Bait was placed inside and when the unsuspecting eagle jumped in to feed it was unable to open its wings to escape. Another more daring but tedious means of capture (Fig. 48) remained in use in Norway until the 1950s (Willgohs, 1961); even today a local ornithologist in the Lofotens uses the method, not to kill but to secure eagles for ringing. A Norwegian called Thome gave an account of the method to Bannerman (1956) which it is worth quoting in full

> Hard by some eagle-haunted fell, the trapper dug a pit about a yard deep, around which he built a rough stone hut of equal height, finally adding a roof of stone slabs. The entrance he closed with a large piece of turf. A little distance away was displayed some carrion, usually a sheep's entrails, secured by a cord which led into the hut. On a favourable morning the trapper would begin his patient vigil before daylight. Sometimes he had not long to wait before an eagle appeared. Even when most sharp-set, the bird would always break off its gorging at intervals to look around for possible dangers. Whenever it was so employed, the trapper carefully drew the carrion a little nearer to the hut. So far from showing any uneasiness as the lure moved, the eagle would tear at it more fiercely. If all went well the bird would follow its meal right up to the hut when the trapper, waiting until his quarry's attention was once more distracted, would seize it by the legs and drag it in. Finding itself in the utter darkness behind the turf curtain, the eagle would offer no resistance. If two or more eagles were at the

Fig. 48. Catching a juvenile Sea Eagle from an 'eagle house', Lofoten Islands, Norway. (Based on a photograph by K.E. Schunemann)

bait together, it sometimes happened that when the first was captured the others continued feeding. With luck a trapper could catch two or three eagles in a few hours. But his patience often went unrewarded, and there he remained crouched in the hut until darkness fell.

On Vaerøy in Lofoten one man caught as many as 16 eagles in a single day! In such favourable localities between 50 and 100 eagles might be caught in any one season. Towards the end of the last century an eagle would be worth 3 Kroner – two as head-money and the third by selling the wings to be used as brooms.

For centuries, then, the White-tailed Sea Eagle had been subject to persecution but this is likely to have been of fairly low intensity or at least local because the species managed to survive with little diminution of numbers. In areas where habitat loss had fragmented the bird's distribution, however, as in southern Britain, persecution even on a small scale would be more effective. By 1800 none bred in England, except for a lone pair which appears to have persisted on the Isle of Man until 1818 (Seebohm, 1883). Many of the Orkney eyries must have been ridiculously accessible locations (Low, 1879) and together with the Galloway nests would appear to have

Persecution and decline

become deserted by this time (Gray, 1871). A single isolated pair on St. Kilda and another on Fair Isle disappeared around 1830 – doubtless suffering, like the Faroese pairs, at the hands of the local cragsman and fowlers.

It was not until the 1840s that a more serious and rapid decline began. Although the celebrated pair at Loch Skeen in Dumfriesshire had gone by 1837, one or two others continued to breed in southern Scotland until about 1866. By this time breeding had ceased on Arran (last recorded date 1849), at the Soutars of Cromarty (1845), Portree on Skye and Lochs Fiag and Fionn in the Northwest Highlands. These were all nests perilously near human habitation or else too easily reached by man, but by the 1860s more remote sites were becoming vulnerable, in such scattered localities as Unst in Shetland, Lee in North Uist, Barra and Glencoe. During the next decade another North Uist site (Baigh Chaise), two in Rannoch (Lochs Ba and Laidon), two others in the Small Isles (Canna and Eigg), one on Mull (Burg) and another on Garbh Island in Sutherland could all be removed from the list. The 1880s saw the abandonment of two or three sites in Assynt, another on Harris and at least two pairs on Skye. The last Irish pair would seem to have bred in 1898, by which time breeding had also ceased on both Fetlar and Bressay in Shetland, but two or three other Shetland pairs (Yell, North Roe and Foula) survived into the twentieth century as did two others in the Outer Hebrides (Ness and the Shiants) and one pair on Rum. A pair had been seen at Ardnamurchan as late as 1913, although breeding was said to have stopped there by 1890, while there are records of a clutch taken in Sutherland in 1901 (Jourdain, in Bannerman, 1956) and of a single Sea Eagle egg reputedly taken on Hoy, Orkney in 1911. The last reported nesting is said to have taken place on Skye in 1916. (For references to these sites see Chapter 3.) Altogether I have located last nesting dates for about 45 White-tailed Sea Eagle sites in Britain. These can be used to illustrate the decline. In constructing the curve in Fig. 49, I have assumed that each site (or its alternatives nearby) was occupied continuously previous to its demise. This is not unreasonable since the Sea Eagle is known to be highly traditional and several sites are expressly described as having been occupied 'from time immemorial'. Successive pairs of Sea Eagles tenanted the Shiants, for example, for 200 years. As each eyrie is said to have been deserted it is removed from our total of 45. Some dates are not sufficiently precise so time periods of 10 years are used. There are, of course, few dates available prior to 1800 but I would propose that up to this time any decrease in White-tailed Sea Eagles was a slow and gradual one, taking place over many centuries. Our curve, if it were taken to be representative, would show that the rate of disappearance of Sea Eagles in North and West Britain gathered momentum from the 1840s until by 1900 only a handful of pairs

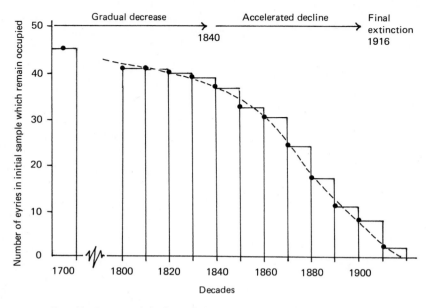

Fig. 49. Suggested decline of the White-tailed Sea Eagle in Britain. For a sample of 45 eyries throughout Britain (excluding Ireland) actual extinction dates have been recorded in the literature. As each became untenanted they have been removed from the total. The histogram shows the number still occupied in each decade and indicates an early period of slow decline when habitat loss and persecution may not have achieved significant proportions, followed by an accelerated decline (seemingly around 1840) until final extinction early in the twentieth century. The resultant curve (superimposed upon the histogram) is remarkably smooth.

were known, all of them in Scotland. Within a further two decades the species had become extinct.

What was it that accelerated the decline of the Sea Eagle when the habitat in the western and northern seaboards of Britain would appear to have changed little in physical appearance? It is known that from about 1550 to 1800 Europe experienced a marked deterioration in climate; with not a little touch of overstatement, geographers often refer to it as 'the Little Ice Age'. Frequent cold winters and wet summers persisted until the 1890s when temperatures at last began to ameliorate (Williamson, 1975). It is unlikely that the White-tailed Sea Eagle could have suffered directly from this climatic change since the species has such a vast geographical range which spans such a diversity of latitude. Indirectly, however, it could have been affected through changes in food supply. One popular theory is that as temperatures rose during the nineteenth century, the range of the

Lumpsucker (*Cyclopterus lumpus*), a fish much favoured by the Sea Eagle, withdrew northwards (D.N. Weir, personal communication). The spectrum of prey which is acceptable to the White-tailed Sea Eagle would seem far too wide for it to be adversely affected by the loss of just one, albeit much favoured, species, however. It has been suggested by Williamson (1975) that warmer ocean currents in the North Atlantic have brought about an increase in other fish species such as Mackerel (*Scomber scrombrus*) and Saithe (*Gadus virens*), the latter especially featuring in the list of prey taken by Sea Eagles. In turn, this has brought about an increase in Gannets (*Sula bassana*), Shags and other seabirds. Thus, even if the Sea Eagle lost one important food species, it might have benefitted from increases in others, whether fish or seabirds. It should of course be borne in mind that some of the changes in seabird numbers might have come about too late to have had any bearing on the Sea Eagle in Britain. Gulls are a notable example. The Eider too has increased in Britain and within a century or so has spread from one or two restricted localities along the entire coastline of the north and west. Its increase is one that the Sea Eagle could and would have made use of to replace any lost prey species. The sharp rise in man's own commercial fishing interests has brought about a decline in stocks of many abundant fish species but again this may have been to the Sea Eagle's advantage since the bird will follow fishing boats to utilise waste thrown overboard.

Competition from other species has sometimes been postulated as one reason for the decline in Britain of the Sea Eagle. Although smaller, the Golden Eagle is more aggressive and there are cases known of its supplanting the White-tailed Sea Eagle from its nest site. Willgohs (1961) has witnessed three such incidents in Norway: one of the displaced pairs abandoned any further attempt to breed that year, but the other two merely moved to alternative nests nearby. It is not common for the two species to have much contact, however, the one being coastal and the other a bird of inland hills and mountains. Although we now find old Erne nests occupied by Golden Eagles, there is ample evidence (see later) to indicate that this is of comparatively recent occurrence, after the Sea Eagle had already long vanished. Some Erne nests have never proved attractive to Golden Eagles and remain vacant to this day.

Other than man, of course, the White-tailed Sea Eagle had no serious predators. Recently, however, it has been shown how the Fulmar can bring about the death of predatory birds such as Peregrine and Sea Eagle (Dennis, 1970), by spitting at them the messy stomach oil. This congeals on the predator's plumage and often proves fatal. It may be tempting to postulate that the meteoric increase in Fulmars in Britain contributed to the decline in Sea Eagles but the latter regularly catch and kill adult Fulmars, presumably in flight, so that only the young chicks or nest ledges may create a threat. One of the Sea Eagles released on Fair Isle as part of an early reintroduction attempt was both young and inexperienced and especially vulnerable. Problems from Fulmars might prove to be very local and of rare occurrence. The Fulmar has increased markedly in recent times but this began at a time when the Sea Eagle was already nearly extinct in Britain; prior to 1878 they bred only on St. Kilda and the subsequent rapid expansion of its breeding range took place after 1900 (Fisher, 1966*b*).

There is little evidence, therefore, to suggest that climate, food supply, or competition from other species had any significant effects on the decline and subsequent demise of the Sea Eagle in Britain. We are left only with human persecution and it is the contention in this book that man alone can be held responsible. His influence may range from unintentional disturbance at breeding sites during critical phases in the breeding cycle (such as incubation) to concerted and wilful destruction of eyries, eggs, young and full-grown birds. The case presented may be a circumstantial one but it is, I believe, nonetheless convincing.

We have seen how the species had endured persecution from man for centuries. Only last century did this reach significant proportions in North and West Britain, because of certain fundamental and abrupt changes in land use. The following account refers only to the Scottish Highlands but is relevant also to Ireland.

The Highlands remained sparsely populated for centuries and it was only during the eighteenth century that the human population began rapidly to increase in numbers. The local economy had been based on beef cattle, with the small native sheep kept mainly for their wool and milk. The stock were herded constantly during the summer months. About 1762 the first Border sheep farmers came north with their hardier breeds of hill sheep and by 1810 huge tracts of grazings were being let out in Caithness, Sutherland and Argyll. In two Sutherland parishes alone the numbers of sheep increased from 7840 in 1790 to 21 000 in 1808; by 1820 there were 130 700 sheep in the whole county (Leigh, 1928). The new husbandry was not adopted in the West Highlands until about 1828, and not in the Outer Hebrides until the 1840s.

On these sheep farms the Highland people were redundant and were 'encouraged' (sometimes forcibly) to settle on the coast where they could gain employment gathering seaweed or fishing (Hunter, 1976). Their number continued to expand and reached a peak by 1841 (Fig. 50); in Orkney and Shetland this peak was not reached for two more decades (O'Dell & Walton, 1962). Thus the increased human population on the coast of the Highlands of Scotland, incidentally in Ireland too (Brody, 1974), put

Persecution and decline

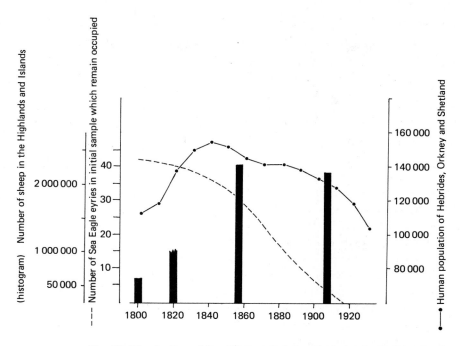

Fig. 50. The decline of the White-tailed Sea Eagle in the Highlands of Scotland in relation to the human population (●—●) (O'Dell & Walton 1962) and the numbers of sheep (histogram) (from Bryden & Houston 1976). The Sea Eagle's decline (taken from Fig. 49) is seen to accelerate at the time of the peak human population on the coast, and at a time when sheep numbers in the area were being increased rapidly.

extra pressure on the White-tailed Sea Eagle, itself of coastal distribution.

From the middle of the nineteenth century the notorious 'clearances' initiated a decline in population, as people were being evicted and transported to the New World, but this allowed no respite for the Sea Eagle. Sheep farming was becoming much less profitable and overstocking had brought about a deterioration in grazing quality. Perhaps seeking a scapegoat for poor husbandry, shepherds harrassed predators like the Sea Eagle with renewed enthusiasm and armed with improved firearms.

Hitherto the flintlock had been efficient only at short range and misfires were frequent. Eagles had to be attracted within range by setting out baits. Macgillivray (1886) described stone huts constructed in the Highlands specially for this purpose. First a pit was dug to a depth of about 1 m and its walls built up with turf and roofed over with sticks and heather. Through a small hole the hunter could sight his barrel on a carcass, or similar bait, placed 30 m distant. One old shepherd was said to have shot five eagles thus in one winter, while another killed three in a single morning (Harvie-Brown & Macpherson, 1904). This was, however, a long and tedious business often for little or no reward.

The instantly igniting percussion cap was introduced early in the nineteenth century and facilitated the shooting of birds in flight. The Sea Eagle possesses an unfortunate habit of circling above its nest, screaming, when someone approaches, presenting an easy target. Often a widowed bird quickly found a new mate – and a fresh victim. A shepherd on the Isle of Rum, for instance, having shot a female at her eyrie, quickly shot her replacement. The male found yet another mate but was himself shot soon afterwards; his widow discreetly abandoned the site. At Fitful Head in Shetland, one of a pair was shot most winters. The other never failed to find a mate until, ultimately, both eagles were destroyed and the eyrie thereafter deserted (Venables & Venables, 1955). The motive for the killing must remain a mystery, for the locals admitted that this pair had never taken lambs.

By far the simplest and most effective means of destruction was to employ poison. Harvie-Brown & Macpherson (1904) mentioned how 40 eagles might gather at a carcass; had this contained strychnine, the consequences would have been devastating. The eagles which came to nest on Loch Ba in Rannoch were poisoned in at least two successive years (Harvie-Brown, 1906). Ussher & Warren (1900) astutely recognised that it had been the use of poisons which had brought about the extinction of the Sea Eagle in Ireland, but understandably we come across few references in the literature to anyone admitting the use of this diabolical mode of slaughtering eagles.

John Wolley was moved to write (1902), 'It is therefore a melancholy reflection that [the White-tailed Eagle] can scarcely exist much longer'. Harvie-Brown & Buckley (1892) went on to plead

> it would be almost endless and not of much account to enumerate the occurrences of White-tailed Eagles recorded all over the districts of our area as shot, trapped or poisoned in past years. What seems infinitely more to the purpose is to 'put up a little prayer' to the proprietor and shooting tenants of lands formerly and presently occupied by White-tailed Eagles to take active measures for their future protection.

One enlightened tenant on the Park estate in the Outer Hebrides had a provision written into his lease (at his own request) prohibiting eagles being killed without express permission from the Chamberlain of Lewis. Perhaps this was deemed necessary to curb the enthusiasm of the estate employees, one of whom had claimed to have killed 13 White-tailed Sea Eagles in one year. Harvie-Brown (1902) recounted how two local men contravened this instruction by taking two eaglets from a nest on the estate and then had the temerity to offer them back for sale to the proprietor. Unfortunately the outcome of this confrontation is not documented!

On Rum, too, the late owner apparently afforded a measure of sanctuary to eagles but by that time they were already rare. 'If they have returned or do return to that island again', wrote Harvie-Brown (Harvie-Brown & Macpherson, 1904), 'the present owner it is hoped, will be as stern and unflinching a protector of them as his father ...' His optimism was premature however, for we know that in 1907, the young heir's game-keeper took a clutch from an eyrie and shot one of the adults. A pair was shot at the same nest 2 years later – all the more regrettable since by then there could not have been more than two or three pairs remaining in the whole of Britain.

In general, eagles posed little nuisance and were tolerated on deer forests, while gamekeepers on grouse moors waged an eternal war on all birds and beasts of prey. As early as 1808, the Marquis of Bute insisted that his keepers swore an oath to 'use their best endeavours to destroy all Birds of Prey, etc. with their nests, etc . . . So help me God' (Richmond, 1959). The Duke of Sutherland offered up to 10 shillings a piece for eggs and adult eagles, while on a neighbouring Caithness estate the same sum was paid for head or talons of an adult, five shillings for a young bird and two-and-sixpence for each egg. On that estate between 1820 and 1826 a total of 295 eagles (of both species) together with 60 eggs or young were destroyed. In the county of Sutherland from 1831 to 1834 a further 171 eagles and 53 young eagles met a similar fate (Harvie-Brown & Buckley, 1887). These figures include both White-tailed and Golden Eagles. During the years 1837 to 1840 the 'vermin' list of Glengarry estate featured 27 Sea Eagles and 15 Golden Eagles (Harvie-Brown, 1904). Not surprisingly, on the coast Sea Eagles were more vulnerable; on Skye, for instance, practically all of the 60 eagles claimed by one keeper had been Sea Eagles (Harvie-Brown & Macpherson, 1904). Of course, one should treat these figures with caution for they are likely to have been inflated, either by the keeper, anxious to demonstrate his diligence, or by the employer, advertising the superiority of his estate management (possibly even by both!). Nonetheless, they do reflect the pressures to which eagles were subjected at that period.

Immatures were more prone to wander than the adults. In a sample of 102 Sea Eagles reported in regional ornithologies for Berwickshire, Northumberland, Yorkshire, Wales, Devon, Cornwall, etc., 80% were immatures and nearly half of them, it is admitted, had been shot. One can presume that most, if not all, were of local British origin and hence added a not insignificent drain on a population already in decline.

Many of these unfortunate birds ended up as trophies in glass cases to adorn Victorian drawing rooms, it being an increasingly fashionable pursuit to collect stuffed specimens and eggshells. Harvie-Brown & Macpherson (1904) were conscious of the extravagant prices offered by collectors. They heard of one notorious egger who had in a period of 25 years, up to 1895, lifted a total of 109 'eagle' eggs; nearly half had originated from only six or seven eyries in Sutherland. One luckless pair contributed 16 eggs over seven seasons. As the White-tailed Sea Eagle became rare it was especially sought after and the exploits of the collectors make villainous reading nowadays; but it should be remembered that they were the naturalists of their time, days when binoculars were unknown and field ornithology was in its infancy. Even the more enlightened Harvie-Brown was not averse to acquiring specimens, but his writings and those of John Wolley and others are now the only record we possess of the White-tailed Sea Eagle living wild in Britain. Even the most ardent protectionist cannot fail to be inspired by Colquhoun's (1888) eloquent and charitable description of a female Sea Eagle at an eyrie in Rannoch; 'her white tail shining like a silver moon . . . When we neared the islet, they both flew out to meet us, uttering their shrill screams. Sometimes they floated at immense height, and then, cleaving the air in their descent, flew round the eyrie, beating their wings which made a hoarse, growling noise . . .' Their nest 'was very deep as well as round. There could not have been less than a cartload of large sticks and twigs'. Peter Robertson told him of a shepherd lad who had robbed the nest in May 1850: 'He has swam in at nicht, the scoondrel, arid ta'en the eggs or young for fear o' his lambs,' adding 'Many a time he has swam Loch Rannoch in the night-time to see his lass' Although pre-empted for this clutch, Colquhoun magnanimously added '. . . when I considered the many night-swims the shepherd had taken for it, I felt glad he had gained his prize, though he had lost me mine'.

Such pairs nesting on islands in freshwater lochs were conveniently accessible and repeatedly plundered. Wolley (1902) surmised how 'the old birds do not always calculate the depth of the water, as there is one place, at least, to which a man can wade'. Others, like our love-struck shepherd, might swim, while one inventive Highlander paddled out in two wooden tubs! The Sea Eagles ceased to breed at the Rannoch site, their eggs being taken, Wolley added indignantly, by a gentleman in a boat. The cliff eyrie at Dunnet Head in Caithness was persistently robbed: two eaglets were shot in 1847, Wolley took another two young the following season and a clutch of eggs in 1849. Not surprisingly, in subsequent years, the adults chose to use a less accessible nest elsewhere. Similarly, the eyrie on Whiten Head in Sutherland is known to have been robbed in 1849, 1852, 1868 and 1874. One wonders in how many other years the eggs or young were taken or destroyed. The situation of this nest, according to Wolley's informant, was awe-inspiring. 'I cannot take upon me to give a description of the wildness of these rocks, only my hair gets strong when I think of them. After being at this place I always felt some dizziness for two or three days.' A similar location was to be found on the towering West Craigs of Hoy in Orkney which rise to a perpendicular height of 400 m from the sea. 'Yet, with the assistance of a short slender rope made of twisted hogs' bristles, did the adventurous climber or rocksman Wooley Thomson traverse the face of this frightful precipice, and for a trifling remuneration brought up the young birds' (Gray, 1871). Such exploits were hazardous in the extreme and the courage of the men who undertook them has to be much admired. One Shetlander climbed up to a nest and actually caught the unsuspecting female Sea Eagle as she incubated. Despite being 'a very expert and daring fowler' the man was forced to drop her on the tricky descent and she was killed (Edmondston, in Saxby, 1874). One of Wolley's climbers took two eggs from a cliff eyrie on Fetlar, but had to drop one while desperately saving his own life. Another Shetlander was less fortunate; when climbing into a nest on Bressay in 1861, 'the poor fellow lost his hold, and of course,' Wolley adds bluntly, 'lost his life'.

In publishing their exploits, Wolley, Colquhoun and Booth were very much in the minority; few other egg collectors ever put pen to paper. All they have left to posterity are faded eggshells gathering dust in drawers and cabinets. Only a proportion of collections are available for reference in museums (and not all of them are now in this country) so that it is difficult to assess the true extent of egg-collecting in the Highlands and its impact on the Sea Eagle. I have amassed data of as many clutches as I can trace, both for Sea Eagles and Golden Eagles. These have been plotted according to the decade of their collection (Fig. 51). There would appear to have been a flurry of activity around the 1850s and 1860s but this could derive from the well-documented activities of John Wolley alone. On the other hand it may be that the Sea Eagle had already gone into decline and its clutches were harder to come by, and as it became scarcer so the efforts of collectors gained in significance and added momentum to the decline. Although shepherds and gamekeepers may have exerted the major impact, collectors at least assisted by mopping up some of the few surviving pairs. From 1900 a staggering number of Golden Eagle clutches found their way into collections. These may suggest a sudden surge in the popularity of egg-collecting but perhaps more realistically reflects both an increased availability of

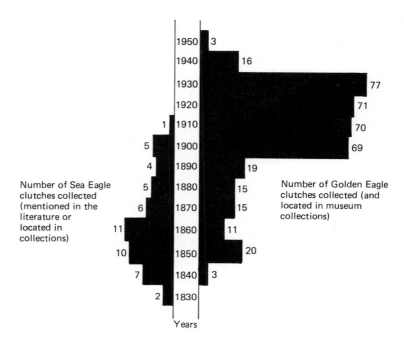

Fig. 51. Egg-collecting and eagles 1830–1950. Relatively few Sea Eagle clutches have been located in collections. In contrast 389 Golden Eagle clutches are known to have been taken, mostly after 1900, perhaps reflecting an increase in Golden Eagles as much as increased collecting efforts – also later clutches will have survived better in collection than earlier ones.

information and an increase in the numbers of Golden Eagles in the Highlands. They contribute a startling 389 clutches altogether, compared with only 51 for the Sea Eagle, although even during peak years this is equivalent to only 7 clutches *per annum* from the whole of the Highland area. Of course, one cannot begin to estimate how many clutches remain to be discovered or have now been lost without trace.

Throughout this account of persecution it is apparent that the Golden Eagle must have suffered heavily, if not as heavily as the Sea Eagle. So why should it be that it survived and still breeds in the Highlands? Several factors contribute to the especial vulnerability of the Sea Eagle. The first is that it too often chose to nest in very accessible locations. We have already seen how researchers in Greenland could gain access to 87% of the eyries they visited: no less than 67% were said to be easily accessible (Christensen, 1979). Similar situations prevail in Iceland (Ingolfsson, 1961) and Norway (Willgohs, 1961) where the habitat shows many similarities to that found in

Persecution and decline

the Highlands and Islands of Scotland. There are no comparable data available for the Golden Eagle but its eyries, located deep in remote mountainous areas, would seem to be much less accessible. To a people concentrated on the coast, the nests of such a large, noisy and conspicuous bird as the Sea Eagle would be especially obvious. Sea Eagles also possess an unfortunate habit of flying overhead, calling anxiously, when there is an intruder near their nest. Not only does this advertise their presence but it also makes them particularly exposed to being shot; Golden Eagles prudently disappear out of sight. It is a rare individual of either species who is prepared to attack a man at its nest: McIan (1848) employed considerable artistic licence in his dramatic painting 'Robbing the eagle's nest' (Fig. 17).

Perhaps the single most important factor contributing to the extermination of the Sea Eagle was its readiness to feed on carrion, and hence its vulnerability to poisoning. This was recognised by Colquhoun (1888)

> A cursory glance will show how much more vulture-shaped both the bill and the body of the Sea-eagle are than those of [the Golden Eagle]. She also partakes of the nature of a vulture in having a less dainty palate than the golden eagle and, being not near so quick a game destroyer, is more apt to devour what she does not strike down. Even carrion does not come amiss, especially in winter . . . she is often fain to content herself with carcasses left upon the inland swamp or cast up by the tide on the shore.

W.H. Hudson (1906) added that this species 'is regarded by the shepherd as the worst enemy to the flock. But the shepherd has his revenge, for the erne is a great lover of carrion and may be easily poisoned.'

Colquhoun reaffirmed how the Sea Eagle was typically a bird of the coast. During the nineteenth century the Highland (and Irish) people also became decidedly coastal in their distribution. In the meantime, inland areas had become depopulated, many becoming deer forest while others had never provided any attraction for human habitation. This was perfect habitat for the Golden Eagle (Fig. 52), indeed Colquhoun and Wolley sometimes referred to it as 'the mountain eagle'. This can neatly be demonstrated by plotting on a map the localities of the Golden Eagle clutches taken by collectors prior to 1916 (the year when the Sea Eagle ultimately became extinct), and comparing it to our distribution map for Sea Eagles (Fig. 53). Only the larger, more mountainous islands of the west, such as Skye, Rum, Mull, Jura, Harris and Lewis, could have supported Golden Eagles, but even here they would seem to have been infrequent. One keeper on Skye had shot many eagles in his lifetime but only one had been a Golden Eagle. Of another 65 known to Harvie-Brown & Macpherson (1904) as having been killed on

Fig. 52. Golden Eagle.

Skye, only three had been Golden Eagles. The Sea Eagle was sometimes found inland, notably the numerous lochs in Sutherland and around Rannoch, but even here they displayed their own distinct preference as the ever-observant Colquhoun (1888) pointed out

> I have enjoyed the rare luxury of seeing both eyries at the same moment, and both queens in undisturbed possession of their thrones. Seldom any collision took place, each having her favourite hunting ground. There was the mountain for the nobler bird, and the morass for her more vulture-shaped neighbour. They sometimes, however, had a battle in the air; but the looser form, the heavier movement and the less daring spirit of the erne made her no match for the mountaineer, who soon drove her screaming to her island...

Her eyrie had been built on a pine tree growing out from the shore of Lochan na h-Achlaise while the Golden Eagles reared their brood halfway up a steep cliff to the east of the loch. Colquhoun added that among the exhibits in the Duke of Sutherland's Museum, the Sea Eagle had been taken as representative of 'the savage rock coast' whilst the Golden Eagle was displayed as a bird of 'the mountain and the deer forest'.

Deerstalking had become a fashionable pursuit in the Highlands. In 1790

Persecution and decline

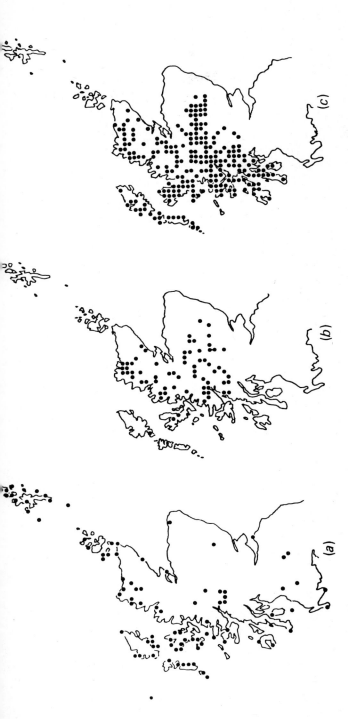

Fig. 53. Distribution of White-tailed Sea Eagles and Golden Eagles in Scotland. (a) Known Sea Eagles eyries. A few Orkney, east coast and lowland eyries date from before the nineteenth century. All had fallen into disuse by 1916. The species was almost entirely coastal except for a few inland sites such as around Rannoch and in Northwest Sutherland. (b) 10 km squares containing Golden Eagle eyries which had been visited by egg-collectors in 1833–1916. They show the eagle to have had an inland distribution while the Sea Eagle inhabited the coast. Other Golden Eagle eyries are known to have existed on some islands such as Skye, Mull, Jura, Rum etc. – all large islands with mountainous terrain. (c) 10 km squares in which Golden Eagles were *proved* to nest during the BTO Atlas survey 1968–72 (data from Sharrock, 1976). At the present time the Golden Eagle has occupied territories on the coast; some of them perhaps old Sea Eagle sites.

there were nine estates used for the purpose and by 1838, when Scrope's famous treatise was published, there were 45. In 1912 the number had risen to 203, totalling some 3 584 916 acres (Bryden & Houston, 1976). That deerstalkers are traditionally more tolerant of eagles than gamekeepers was cleverly demonstrated by Sandeman (1957a). He found that the proportions of lone eagles on territory, or of one partner of a pair being immature, were greater on sheep ground and grousemoors than in deer forests. This was the direct result of persecution. The number of chicks actually fledging on sheep ground and grousemoors was only 0.3 per nesting attempt, compared with 0.6 on deer forests. Thus we can comprehend why the localities of our Golden Eagle clutches from last century should correspond so closely with areas exploited as deer forests (Fig. 54). During the worst phases of persecution, Golden Eagles probably derived ample sanctuaries from which to stage a comeback. They gained respite when gamekeepers were employed in military duties abroad during the First World War, but by this time the Sea Eagle was extinct.

The depressive effects which can result from human persecution have been exhibited by other raptor species. Many other species declined and seven had disappeared from Scotland early this century – the Sea Eagle, the Red Kite (*Milvus milvus*), the Osprey, the Goshawk, the Hobby (*Falco subbuteo*), the Marsh Harrier (*Circus aeruginosus*) and Honey Buzzard (*Pernis apivorus*) (Newton, 1972). It may be significant that the Red Kite displays a fondness for carrion and like the Sea Eagle this may have played a significant part in its ultimate downfall. The Air Ministry's harassment of Peregrine from 1940 to 1945 virtually eliminated the species from the southwest of Britain, and considerably reduced it elsewhere. Once the persecution was relaxed the Peregrine recovered its former abundance within a decade or so (Ratcliffe, 1980). Newton (1972) has been able to show how the Sparrowhawk was able to recover its numbers during the Second World War while persecution was reduced to a minimum. Buzzards had also been depleted but were able to build up numbers only in those areas of Britain where gamekeepers were scarce or least active (Moore, 1955).

In summary, it is probable that the White-tailed Sea Eagle was widespread throughout the British Isles. From Anglo-Saxon times – about 1000 years ago – its habitat in lowland Britain began to diminish. It also had to endure a measure of persecution by man but this seems to have been effective only where the species was already scarce or its habitat had been fragmented. The remote coasts of Ireland and northern Scotland could easily have maintained a viable population of the species but several factors came into play during the nineteenth century which were to pose a serious threat. The human population increased and suddenly became more

Persecution and decline

Fig. 54. Distribution of Golden Eagles in Scotland late last century, in relation to deer forest. The shaded regions are areas of deer forest in 1883 (taken from O'Dell & Walton, 1962). The dots represent 10 km squares (National Grid) in which a Golden Eagle eyrie was known to have been robbed in 1833–1916.

coastal so that interactions between man and Sea Eagle became critical. The bird came into conflict first with sheep farming interests and later with sporting interests. As shepherds and keepers reduced its numbers, it became a popular target for egg- and skin-collectors. Large birds of prey are particularly susceptible to persecution because of their long life but slow rate of reproduction. The White-tailed Sea Eagle was particularly susceptible to depredations by man because of the accessibility of its nest sites and the comparative ease with which it could be shot at the nest or poisoned at carrion. In contrast, the Golden Eagle gained a measure of protection from its inland habitat (where the human population was scarce) and by being tolerated wherever this was used as deer forest. It survived in sufficient numbers to gain respite from the two World Wars and to increase subsequently. Indeed it now occupies some former Sea Eagle territories on the coast.

It would appear from endless examples, only some of which are reviewed, that the extent of human persecution directed towards the White-tailed Sea Eagle (amongst other raptors) was sufficiently intense to result in its extermination in Britain. A similar sequence of events occurred in Europe where the species is also much threatened. Some of the steps being taken to save it will be discussed in the next chapter.

7 Conservation

> Our natural sense of regret at the extermination of the [sea] eagle in Skye is tempered by the consideration that the wide range it enjoys renders it almost impossible that it could ever become a lost species, since it ranges from Greenland to Siberia, and from the fjords of Norway to the forests of Hungary.
>
> Harvie-Brown & Macpherson (1904)

Alas, despite Harvie-Brown and Macpherson's complacency, the White-tailed Sea Eagle is very much a threatened species. Without doubt, even as they wrote the bird's habitat was rapidly diminishing at the hands of man. The forests of Hungary, as elsewhere on the continent, were being felled at an alarming rate. The wetlands of Europe were being continually drained as new farmland was required to feed the ever-increasing human population. Even in northern latitudes where habitat loss was less significant, eagles were subject to massive persecution: we have already looked at examples of this but an exhaustive, and depressing, review can be found in Bijleveld's (1974) *Birds of prey in Europe*. Collecting too was becoming increasingly popular and Bijleveld tells how in only two decades one notorious dealer was said to have handled over 400 Sea Eagles which had been shot in Romania and how, during a 12-day expedition to the Yugoslav Danube in the spring of 1878, Crown Prince Rudolf shot 12 Sea Eagles along with 37 other raptors of 11 different species. In one part of Germany in 1852 one man had robbed four of the only five Sea Eagle eyries to have been occupied that year.

To habitat loss and persecution we now have to add a third threat – contamination by toxic chemicals: this will be discussed shortly.

Together, these three factors have brought about a drastic decline amongst White-tailed Sea Eagles in almost every country of Europe: in some the species has ceased to breed altogether. In order to halt and to reverse this lamentable state of affairs several conservation measures have been implemented.

The earliest efforts were directed at the birds themselves, laws being passed to prohibit their wanton destruction. This was initiated as early as 1913 in Iceland where the Sea Eagle was on the verge of extinction. Sweden and Finland followed suit in the 1920s and Denmark in 1931. The laws brought into effect in Germany were complex to say the least. In Prussia, for instance, it was permitted to kill Sea Eagles and Ospreys but *not* by using firearms. The Sea Eagle was accorded legal protection in Pomerania in 1922 and in Mecklenburg 4 years later but not until 1935 did protective laws extend throughout what is now East Germany. The 1935 Act in Germany decreed that all birds of prey were game but conveniently omitted to specify an open season when they might legally be shot! Other European countries were slower to act, some not until the 1950s or 1960s. Despite active campaigning by ornithologists, Norwegian eagles did not achieve protection until 1968. Three years later eagles became protected in Italy and in 1973 in Greece. Both Bijleveld (1974) and Conder (1977) review the current legal status of all European raptors.

It is one matter to accord legal protection but it is another to enforce it: illegal shooting is still a problem. Between 1946 and 1972 in East Germany for instance, a total of 194 Sea Eagles were picked up dead – nearly half of them had been illegally shot (Oehme, 1977). In Greenland, laws were reluctantly passed much against the wishes of the local sheep farmers (Wille, 1977). Similarly, in Norway controversy arose about the prospective damage to livestock but rarely did the government entertain compensation claims: indeed few of those 8000 claims investigated by Hagen (1976) were found to hold any foundation. He was gratified to discover instead that 60% of Norwegian principalities were happy that eagles remain protected!

In some countries, such as Iceland and Greenland, Sea Eagles continued to be destroyed in traps set for Arctic Foxes. Being partial to carrion, the Sea Eagle is easily poisoned by baits allegedly laid to kill other species. In Bulgaria, hundreds of wolves are poisoned each year with some 60 kg of strychnine being employed annually. The situation was even worse in Romania but here, mercifully, the practice has become illegal (Bijleveld, 1974). An ironic twist to the story is that even should eagles escape direct poisoning themselves, the loss to them of carrion made available as wolf kills seems to be detrimental.

An alternative approach to conservation of the Sea Eagle is to prevent further degradation or fragmentation of habitat. Legislation has resulted in areas of prime habitat in Norway and Sweden, for instance, being accorded Nature Reserve status. Over 1% of the total land area of Mecklenburg is now contained within such reserves: 11 of them together hold some 14 breeding territories of Sea Eagles (Oehme, 1969a). In addition, since 1967 the World Wildlife Fund's EUREL scheme has attempted to persuade private landowners to manage their property to the benefit of wildlife. These powers have been employed to good effect in Schleswig-Holstein, for example, where in 1970 75% of the potential and current Sea Eagle territories were thus protected (Bijleveld, 1974). The Department for Nature Protection and Landscape Care holds powers to declare areas around Sea Eagle eyrie trees a sanctuary to limit public access and also to forbid trees being felled in their vicinity. Similarly, in East Germany the felling of nest trees became illegal in 1955, and in 1961–63 all timber operations within 300 m of a nest tree ceased during the breeding season: only wind-blown and dead trees can be removed within 100 m of the eyrie at any time of the year (Oehme, 1969a). The Swedish Forest Service limits logging activities to within 200 m of an eyrie tree and during the breeding season this safeguard is extended to a radius of 1000 m. However, to private forest owners the Service can only act in an advisory capacity so that its recommendations are not compulsory (Helander, 1975). In Poland, some 78% of forests are state owned and clusters of 10 or more trees are preserved around each Sea Eagle eyrie (Bogucki, 1977). The National Board of Forestry in Finland similarly curtails timber activities on state property where eagles, Ospreys and Eagle Owls nest: unfortunately no Finnish Sea Eagles have yet recognised the advantages of nesting on such state ground. All nest trees are, however, marked by a sign explaining that they are protected by a nature conservation law. On one island in Finland the World Wildlife Fund has actually purchased all the timber so that two Sea Eagle eyries can be safeguarded for the future (Joutsamo & Koivusaari, 1977). Similarly, in Schleswig-Holstein, compensation has been paid for the long-term preservation of mature timber where two pairs of Sea Eagles breed (Ruger, 1981).

Since the early 1960s, East Germany has operated a reward scheme of 100 Marks for every brood of Sea Eagles fledging successfully on private property (Oehme, 1969a; Bijleveld, 1974): a system similar to that once organised in Britain by the Royal Society for the Protection of Birds for Golden Eagles.

Nesting territories merit protection not only from man's commercial designs but also from his leisure pursuits. In both Sweden and Schleswig-Holstein, there is political pressure to open up, for recreational use, many of the lakes in Sea Eagle breeding habitats; while in Greenland the building of lakeside summer houses can pose a threat. In East Germany, restrictions are placed on hunting near Sea Eagle eyries and public access is limited. Even in remote areas, like Kandalaksha in Arctic USSR, human interference can be a problem (Flerov, 1970). Eyries in Swedish Lappland, which are especially vulnerable in this respect, enjoy special protection (Helander, 1981a). Some countries, such as Iceland, forbid photography at Sea Eagle nests. Even improper and persistent visits by well-meaning ornithologists can cause desertion: as Oehme (1969a) observed in East Germany, 'Sea Eagles no longer have enemies here – now we must protect them from their friends'. Each occupied eyrie in Schleswig-Holstein has enjoyed since 1974 a round-the-clock guard by teams of German and Dutch volunteers, over 100 being recruited each season. At the foot of each nest tree a sensitive microphone is positioned, wired to a loudspeaker in the observation caravan some 200 m distant (Fig. 55). All nest trees, including the alternative sites not in use that year, are surrounded by barbed wire.

Implicit in all these measures is a well-informed and educated public in whom a sympathy towards birds of prey and their conservation should be encouraged.

Having accorded Sea Eagles legal protection and taken steps to safeguard both their habitat and their nests, a further measure may be contemplated aimed at increasing the breeding density. Saurola (1978) considered that only 0.1% of trees within Sea Eagle habitat in Finland may be suitable for a nest. Artificial eyries have been constructed in areas where it was known that eagles have bred in the past. Nineteen such nests were made available by the late 1970s in the Quarken region and in north Finland: 14 have been visited by White-tailed Sea Eagles, eggs have been laid in at least nine of them, and two ultimately fledged young (Stjernberg, 1981). In Sweden, three such artificial nests have been adopted by Sea Eagles for several breeding seasons in succession (Helander, 1981a). This technique is used widely in North America for Ospreys and Bald Eagles, some of the latter species willing to nest on platforms constructed atop electricity pylons (Nelson, 1978).

Any such conservation measures seem futile, however, if the eagles being protected are repeatedly unable to breed successfully because of toxic chemicals in their environment. DDT has been in widespread use as an insecticide since the late 1940s. Another organochlorine, dieldrin, came into use as a sheep dip, while alkyl mercury compounds are employed as seed dressings against fungi, etc. In the late 1960s, polychlorinated biphenyls (PCBs) were identified as a biproduct abundant in industrial wastes. The stability and persistence of these chemicals result in their accumulation at increasing dosages at each level in the food chain.

Fig. 55. A typical West German eyrie on top of a 30 m-high beech tree: during the breeding season it is under constant guard from an observation caravan some distance away, Schleswig-Holstein, March 1977. (Photo: J.A. Love)

Predators such as the Peregrine and Sparrowhawk, at the top of the food chain, accumulate amounts sufficient to impair their breeding if not actually to prove lethal. Newton (1979) has reviewed the effects of pollutants on raptors, while Ratcliffe (1980) pioneered detailed study of the problem amongst Peregrines. Once the use of toxic chemicals was restricted in Britain, Peregrines recovered their numbers and breeding success improved. DDT was still leaching from agricultural land into the sea, however, and falcons nesting on the coast, where they preyed upon seabirds, were slow to recover.

Being an aquatic and coastal predator, the White-tailed Sea Eagle is immediately vulnerable, especially where it chooses to frequent shallow or inland seas such as the Baltic, or the freshwater lakes in agricultural areas. In the early 1960s in Finland it was noted that Sea Eagles were experiencing a much reduced breeding success. In one area during the 1966 season, no eggs were laid at all and eight full-grown birds were found dead. An analysis of the corpses revealed concentrations of mercury which could have been sufficient to be fatal (Henriksson *et al.*, 1966). Further studies showed exceedingly high levels of DDE/DDT (25 000 ppm in lipid) and PCBs (13 000 ppm), not only in Sea Eagles but also in their eggs (1000 ppm DDE and 600 ppm PCBs) – the highest levels recorded in any raptor eggs (Jensen, Johnels, Olsson & Westermark, 1972). Many such eggs prove infertile, whilst DDE is also known to reduce the thickness of the shell making it more liable to damage. Normal eggs in Finnish museums, collected between the years 1884 and 1935, previous to the DDT era, were found to have an average shell thickness of 0.62 mm; a sample of 20 laid between 1969 and 1975 averaged only 0.52 mm, a reduction of 16% (Joutsamo & Koivusaari, 1977).

Museum specimens have also proved useful in assessing the effects of heavy metals. Some natural mercury has always existed in the environment and in a study of Swedish birds Berg, Johnels, Sjöstrand & Westermark (1966) found, not surprisingly, that the highest concentrations occurred in predatory species. Museum skins of Peregrines contained about 2500 mg/g and Sea Eagles as much as 6600 mg/g: smaller quantities were detected in Partridges and pheasants (only 100–280 mg/g). Following the introduction of alkyl mercurial seed dressings in the 1940s there occurred a marked increase in detectable levels of mercury – up to 60 000 mg/g in Sea Eagles. Residues have since remained high, Sea Eagles and their eggs having the dubious distinction of containing the highest amounts recorded in any Swedish wildlife (Borg, Wanntorp, Erne & Hanko, 1965).

It is the birds nesting along the Baltic coasts which are the most affected in Sweden, while those breeding in Lappland have only minimal amounts of mercury, DDT or PCBs (Jensen, Johnels, Olsson & Otterlind, 1969; Helander, 1975). There has been a steady decrease in the number of pairs breeding in Baltic Sweden while their eggshell thickness has diminished by some 11–17% since DDT came into use (Helander, 1975).

Toxic levels have been examined in carcasses of adult eagles and eggs from West Germany (Koeman, Hadderingh & Bijleveld, 1972; Ruger, 1975). One adult contained 48.2 ppm mercury in its liver and 2615 ppm in the kidney, levels which have proved lethal to raptors held in experimental conditions (Borg, Erne, Hanko & Wanntorp, 1970; Koeman, de Goeij, Garssen-Hoekstra & Pels, 1971). Such contaminants have also been discovered in Pike and in waterbirds such as Coot, which are the prey of Sea Eagles in West Germany (Koeman et al., 1972). Heavy pesticide applications are made across the border in East Germany where Rape is cultivated extensively, its oil being used in margarine manufacture; in 1976 T. Neumann knew of only one successful Sea Eagle nest out of 28 visited. Oehme (1969b) has documented a marked increase in poisoned Sea Eagles amongst those deaths being reported, from 6.4% during the years 1946 to 1957, to 24.6% from 1958 to 1965. In 1964 alone, 20 eagles were found dead; five deaths could be attributed to parathion used in crow control and others were suspected of having fallen victim to agricultural pesticides. Information is lacking from elsewhere in Europe but Bijleveld (1974) reported that from 1957 to 1967 pesticides had reduced the Sea Eagle population in Hungary by 50–60%. With the alarming state of affairs existing in the Baltic area, there is an urgent need to investigate other inland seas such as the Caspian and the Black Sea which have both breeding pairs of Sea Eagles and wintering birds from further north. The last two pairs of Sea Eagles to nest in Israel disappeared during the early 1950s when farmers began freely to apply thalium sulphate and chlorinated hydrocarbons such as DDT (Y. Yom-Tov, personal communication).

Unless the level of contamination is so high as to cause death, the principal route through which toxic chemicals affect birds of prey is by lowering their breeding output. About 75% of nesting attempts in West Germany, Finland and the Swedish Baltic ultimately fail, and only some 0.3 young are reared per occupied nest. It has been demonstrated that pesticide levels are low in Swedish Lappland where the breeding success of Sea Eagles is accordingly high – about 70% of nests succeeding and on average 1.1 young were produced from each occupied nest (Table 12): this is similar to the reproductive output in the uncontaminated populations of Norway. It is not easy to assess what level of breeding output needs to be attained to

Table 12. *Productivity in White-tailed Sea Eagles in Scandinavia and Germany*

Country	Years	Percentage of nests which were successful	No. of young per successful nest	No. of young per occupied nest	Source
Norway	1956–60	60.0	1.6	c.1	Willgohs, 1961
	1974–76	c.60	1.6	0.9	Norderhaug, 1977
Sweden					
Lappland	1964–76	71.0	1.5	1.1	Helander, 1975, 1977
Baltic	1970–76	23.0	1.3	0.3	Helander, 1975, 1977
Finland	1970–80	22.0	1.3	0.3	Stjernberg, 1981
West Germany	1948–56	60.0	1.14	0.7	Fischer, 1970
	1964–76	26.0	1.3	0.3	Neumann (in Fentzloff, 1977) and personal communication)
East Germany	1953–54	73.0	1.47	1.1	Fischer, 1970

maintain a healthy and viable population but for Bald Eagles Sprunt *et al* (1973) have estimated that some 50% of nesting attempts should succeed or at least 0.7 young be reared per occupied nest.

In his important book *Havsörnen i Sverige*, Bjorn Helander (1975) wrote:

> The prerequisites for the survival of the sea eagle in Sweden are protection of the breeding areas and a reduction in pesticide levels in the environment. Pesticide pollution is an international problem, and international restrictions on the use of toxic substances are necessary. It is our duty to try and save the White-tailed Sea Eagle and its habitat until the environment is again clean enough to maintain healthy populations.

It appears that terrestrial situations may register some recovery relatively quickly but the sea, being the ultimate sump for pollutants, takes very much longer. Goldberg *et al.*, (1971) made the gloomy prediction that only 25% of the organochlorines already dispensed has yet reached the sea, and it may be decades before their levels are reduced even by 50%. If so, Sea Eagles are likely to experience a depressed breeding output for many years to come. There are few populations which are not under threat in some guise or other, so that immigration of surplus birds from healthy populations is unlikely to be an adequate alternative. Few options remain: one is to protect stringently the depleted habitat and its handicapped inhabitants until such time as the situation might improve; a second is to attempt to reduce the contamination in the breeding population; a third is to enhance the survival of those few young which do reach maturity; and a fourth is to supplement the output of young by artificial means.

The first option may be exercised, as we discussed earlier, by the setting up of Nature Reserves in suitable habitat, the provision of additional artificial nest sites, legislative protection of the breeding population from human interference and persecution, and the guarding of nests.

The second and third options are presently being attempted by providing supplementary uncontaminated food. This may enhance the survival and breeding capabilities of the adults, by reducing the concentrations of toxins within the birds to acceptable levels. In Sweden are used whole carcasses of road-killed or domestic animals, together with offal from slaughter houses (much of this obtained free of charge). Offal is easily carried off by crows, etc., and rarely lasts more than a day or two, while an equivalent weight of carcasses will remain for up to a week. The Swedish Wildlife Service (SNF) co-ordinates a nationwide team of volunteers to help in the replenishment of the food supplies. The food is placed in situations which are free from human disturbance and where eagles can enjoy a clear view of their surroundings. Frozen lakes or inlets, fields, bogs and marshes are most suitable. Special elevated platforms are constructed where foxes might be a nuisance in carrying away food. A continuous store is provided from October to March; thereafter the eagles prefer to catch live prey. In the winter of 1976/77, for example, 80 tons of food were provided at 96 feeding stations (Fig. 56) throughout central and southern Sweden (Helander, 1975, 1978b).

A similar programme was initiated along the southwest and west coasts of Finland in 1972 and by 1978 some 31 feeding places were in operation (Hario, 1981). Winter feeding was intended first to reduce the contamination in adult Sea Eagles but it is difficult to assess any resultant improvement in breeding success. Data for 12 pairs in Sweden showed that breeding success improved from 29% to 44% with the introduction of feeding. Two of those pairs did not succeed at all during the entire period; the improvement shown by the remaining pairs – from 36 to 56% – was statistically significant ($p < 0.05$) (Helander, 1981b). It can be argued that any further accumulation of toxins in the Swedish and Finnish populations has been halted, and that the clean food is maintaining the level of reproduction of the less seriously affected pairs. If the degree of pollution diminishes in the near future the scheme could also hasten the process of recovery in nesting success (Helander, 1978b).

More and more juveniles are utilising the food dumps each year, which

Fig. 56. An adult, sub-adult (right foreground) and immature (behind) White-tailed Sea Eagle at a food dump in southern Sweden. (Photo: B. Helander)

indicates that those being reared in the vicinity are experiencing an enhanced survival rate – both because of the extra rations available to them at a time when it may otherwise be difficult for them to find alternative prey, and because this ready supply of food is deterring them from moving away for the winter months. To disperse south is a hazardous undertaking, with a high risk of illegal persecution on the way or in the winter quarters. The first signs of improved recruitment to the Swedish population was detected in 1978, when five of 11 successful pairs contained one bird that bred for the first time, and two previously unoccupied, old territories were occupied by immature pairs (Helander, 1981b; Hario, 1981).

The fourth option, to supplement the reproductive output of a contaminated population, can be achieved by several different means. One is to provide pairs which have repeatedly failed to hatch any eggs with eggs taken from pairs breeding in uncontaminated areas. As far as I am aware this has not been undertaken with Sea Eagles; in 1974 such transplants were made with Bald Eagles in Maine, using 'clean' eggs from Minnesota, but it was found that two eggs which had been removed from a contaminated pair in Maine later hatched successfully in an incubator. The young were subsequently fostered back into a wild nest but this whole incident, despite its successful outcome, attracted some adverse criticism and the technique has not been attempted again (P. Nye, personal communication; S. Postupalski, personal communication).

A second method devised to augment the breeding output of threatened populations is to release eaglets bred from captive pairs. This can be a tricky and expensive achievement but one which is meeting with more and more success each year, especially with Bald Eagles. Reviewing the captive propagation of Bald Eagles, Hancock (1973) knew of 20 young which were known to have been so raised; this was equivalent to an 83% fledging success and an average of 1.9 young reared per successful nest. This compares favourably with even healthy wild populations (for which Hancock quoted a 57% success rate and 1.9 young per successful nest among wild Bald Eagles from British Columbia). Maestrelli & Wiemeyer (1975) described additional successes at the Patuxent Wildlife Research Centre in Maryland but clearly with over 130 Bald Eagles held in captivity throughout the USA (Thacker, 1971) the full potential has yet to be realised.

The very first successful breeding of captive Bald Eagles occurred in Ohio in 1886 (Hancock, 1973) but it was not until 1961 that a White-tailed Sea Eagle chick was raised. This took place in Vienna Zoo (Fiedler, 1970) where both its parents had been acquired as juveniles in 1955. They were kept in a large aviary with various other eagles and vultures and were often seen to indulge in courtship behaviour. In 1961 they suddenly produced an egg and vigorously defended their corner of the aviary once the egg had hatched, actually killing two intruding Bearded Vultures (*Gypaetos barbatus*). The chick was eventually removed and hand-reared. The following December courtship began again so the pair were rehoused in a smaller aviary nearby. They laid two eggs which both hatched but only one chick was raised. Single chicks were reared each succeeding year until 1968, when the eaglet died. As if to compensate for this failure, discreet human intervention induced two young to fledge the following year but thereafter, possibly due to very cold spring weather, no further clutches have hatched (Fiedler, 1977).

A pair of Sea Eagles has been kept in Tel-Aviv Zoo (Prof. H. Mendelssohn and Dr Y. Yom-Tov, personal communication) and in 1976 the female laid a clutch of two eggs from which one chick fledged. Their aviary (Fig. 57) measures $4 \text{ m} \times 6 \text{ m}$ and is 4 m high, with the roof and three of its walls made of 2.5 cm wire mesh. The nest is constructed on a square platform in the centre of the cage (Y. Yom-Tov, personal communication). Each subsequent year the female has laid clutches of two or three eggs from which one, two or even three young have been reared (Table 13). From the 15 eggs laid over six seasons, 10 young (67% success) have fledged – a

Table 13. *Breeding of White-tailed Sea Eagles at the Wildlife Research Centre, Tel-Aviv University (after Prof. H. Mendelssohn)*

Year	Date of first egg	No. of eggs in clutch	No. of young fledged	Remarks
1976	13th Feb.	2	1 ♂	2nd egg failed to hatch
1977	30th Jan.	3	1 ♂	2 eggs infertile
1978	16th Jan.	2	1 ♀	2nd egg broken
1979	23rd Jan.	3	3 ♀ ♀ ♀	
1980	24th Jan.	2	2 ♂ ♂	
1981	29th Jan.	3	2	3rd young died when nest collapsed

Mean ($\pm$SD) 28th Jan. 2.5 ($\pm$0.5) 1.8 ($\pm$1.0)

Fig. 57. White-tailed Sea Eagle breeding cage in Tel-Aviv Zoo, Israel. The nest lies in the box at the top right-hand corner of the cage where the mirror allows its contents to be viewed without entering the cage. A turtle basks in the pool. (Photo: Prof. H. Mendelssohn).

remarkable degree of success from a captive pair. Professor Mendelssohn intends to utilise the progeny to reintroduce the White-tailed Sea Eagle to the Jordan Valley in northern Israel. Already (1981) a male (from 1976) and a female (bred in 1978) are together in a cage on the site awaiting release.

A pair of Sea Eagles has been kept for some years in a vast aviary at Eekholt Wildlife Park in Schleswig-Holstein. In 1977 the female laid two eggs but despite incubating them for 58 days they were infertile. That same season she laid a replacement clutch from which was reared one eaglet (Brüll, personal communication); young have been reared in subsequent years.

Consistent successes have been achieved at Burg Guttenburg in West Germany where Claus Fentzloff (1977, 1978) has two pairs now breeding annually with the prospect of other pairs soon to join them. The first pair was made up of a female, Clara, then 20 years old and her 10-year-old mate Korsar. They were put together in March 1971. Their aviary measured 6 × 8 m and was 5 m high. Clara did not accept her new mate at first, so Fentzloff tethered them within sight of one another. Korsar, who was trained to the fist, was flown above Clara's perch whenever possible; Clara herself had once been injured and could not fly. She began to show more interest and when they were put together in December 1971, she finally accepted him. Within 2 months they had begun nest-building and on 13th March 1972 one egg was laid. Both adults took part in incubation but broke the egg 12 days later.

Nest-building began anew the following February, the first egg appeared on 2nd March and another 2 days later. After about 10 days' incubation, one egg was found to be broken, so the second was removed to be incubated artificially. Fentzloff wondered whether Clara's eggs might be abnormally thin shelled so in the interim he fed the pair additional calcium, phosphorus, trace elements and vitamins. On 1st April, 19 days after the loss of her first clutch, Clara laid a replacement egg which broke almost immediately before she produced a fourth. This was immediately removed but the parents were kept broody by being allowed to sit on a substitute goose egg. The surviving egg from the first clutch was accidentally damaged in the incubator and the fourth egg proved to be infertile. After 48 days Clara and Korsar finally gave up brooding their goose egg. In 1974, two eggs were again laid at the end of February but recurring problems with the incubator resulted in the death of the embryos. Clara's single replacement egg, laid on 29th March, proved infertile.

Undaunted, Fentzloff prepared for the 1975 season by improving his incubator and in May 1974 moved the eagle pair to a new and larger aviary. Its solid walls measured 9 × 13 m × 5 m high. It had bars near the roof and the roof itself was covered in 10 × 20 cm wire mesh. This 'skylight and seclusion' enclosure had been employed so successfully by Hurrell (1977) for a variety of other raptors. During October and November 1974, preliminary courtship was observed and nest-building began the following January. The lining of the nest was completed by 12th February in readiness for the first egg, laid 10 days later; the second was laid on 1st March. Clara and Korsar were permitted to complete incubation on their own and both eggs hatched, each 38 days after being laid. Both chicks were reared to fledging. This success was repeated in 1976 when two more eaglets were raised and when a second pair (6-year-old Pirat and 20-year-old Thora) also produced its first clutch. The eggs were laid on 29th February and 3rd March, in a scrape on the ground. After 9 days, Fentzloff removed them to his incubator but both proved infertile. In the meantime Thora had accepted a substitute goose egg which she happily incubated in an upraised artificial nest which Fentzloff had constructed for her. These two eagle pairs produce five eggs during the 1977 season from which two chicks were hand-reared and a third was brought up by Clara and Korsar. Such successes have been repeated in subsequent years.

No sooner had Fentzloff produced the first captive-bred eaglets than he began the first of his fostering experiments. The application of this technique was limited as few wild nests were available in Schleswig-Holstein and even fewer proved suitable. The two eaglets reared in 1975 were both fostered back on to wild pairs. On 24th April, a 19-day-old captive-bred chick was put into the eyrie of a pair which had lost its own clutch of eggs. The adults were slow to return to the nest and because of the cold weather the chick was removed again to be brooded overnight. The adults again refused to accept it the next day and that attempt had to be abandoned. Another pair which already had a chick of its own was chosen for a second attempt but this was not made until the captive-bred eaglet was much older, at 7 weeks. The wild eaglet at first showed aggression to its new sibling but soon grew to accept it. When the adults returned with prey, 9 hours later, they first fed only their own chick; 15 minutes later, when the newcomer begged, one of the adults moved across to feed it also. Three or 4 weeks later, both eaglets fledged successfully. A second separate adoption was made with the other eaglet captive-bred during 1975. It too was readily accepted by a wild pair but when both chicks fledged 5 weeks later, the wild chick was found to be incapable of flight. It was thought that a high infestation of flagellate protozoans had caused abnormal feather development and at first fears were expressed that the condition had been passed to it by its captive-reared nest mate. Veterinary investigation later showed this to be unlikely.

None of the eyries in Schleswig-Holstein proved suitable for adoptions in

1976. Only one hatched any eggs and both chicks were reared to fledging, but Fentzloff became involved in an attempt to save a clutch of three eggs. The female had been found apparently suffering from the effects of poisons and was taken into captivity for treatment: she later recovered and was released. Meanwhile her eggs were incubated artificially but only one hatched. At 50 days of age it was fostered on to Clara and Korsar, their own two well-grown young being temporarily removed. The hand-reared eaglet was much younger but the adults were stimulated to feed it by a tape-recording of food-begging calls being played nearby. All progressed well and in order that the chick would first become used to nest-mates the two older eaglets were returned to their parents. On 30th June it was decided to foster the chick back into the wild but there were no suitable nests in Schleswig-Holstein so Bjorn Helander in Sweden was contacted. The chick was taken by air to Sweden where Helander had selected an eyrie with an eaglet of similar age to the captive-bred one; the parents readily accepted this new addition to their family. The foster twins fledged successfully by 8th August (Helander, 1976).

The precariously small West German population suffered further setbacks in 1977. One pair abandoned its breeding attempt but its clutch of two eggs was rescued by the local conservation body. Both eggs hatched and the young were given to the Fentzloffs. Clara and Korsar were already busy with a brood of their own but Pirat and Thora had abandoned their ill-fated breeding attempt. They were restimulated using a goose egg and shortly afterwards had it replaced with a young buzzard. This too they nonchalantly accepted, indicating to Fentzloff that they would also take one of the young hand-reared eaglets instead. This they did until such time that it was old enough for being fostered back into the eyrie of a wild pair. The pair chosen already had a chick of its own but despite its being 17 days older than the foster chick, both were tended by the adults until they fledged in July. The second eaglet hatched from the rescued clutch was found to have a deformed oesophagus but responded to veterinary treatment. Once given a clear bill of health it was ready to be introduced back to the wild. This was achieved by constructing an artificial nest platform in the territory of a pair whose breeding attempt had failed earlier that season. The eaglet was placed within and fed each day by hand until it flew on 16th August. The territory holders took immediate interest and were often seen in the company of this strange youngster: on 13th September they were seen to provide it with food! This may at first seem remarkable but I have witnessed similar behaviour in sub-adults with juveniles released on Rum. These and subsequent events in the small West German populations are summarised in Fig. 58.

Sea Eagles seem amenable to such new dimensions in the conservation of

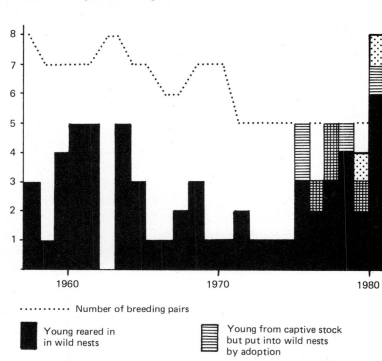

Fig. 58. Number of White-tailed Sea Eagles breeding in Schleswig-Holstein, West Germany, and their annual breeding outputs (latterly augmented by artificial means). (Data from Ruger, 1981).

the species. In 1978, the Swedes undertook an elegant experiment in cross-fostering. Three of their Baltic pairs had, since 1964, repeatedly failed to rear any young, so, early in incubation, their eggs were removed. Two of the clutches proved infertile but from the third, one chick hatched. It was hand-reared for 35 days and weighed 4.2 kg when it was introduced back into a wild nest. This nest contained three young. Two of them were of similar weight to the hand-reared chick which had been put in beside them, but the youngest weighed only 1.9 kg. This 'runt' was removed and fed for 2 days before an attempt was made to introduce it into the nest of a second wild pair which were vainly incubating an infertile clutch. The adults immediately deserted, however, so the chick was rescued and fed for a further 3 days by which time it weighed 2.3 kg. At last it was accepted by a third pair of eagles that had two chicks of similar weight, though a week younger than their foster chick; all three fledged in July (Helander, 1979).

A similar operation had been carried out in USA in June 1974, with two 9-week-old Bald Eagles rescued from a collapsed eyrie in Michigan. Both were adopted by two other wild pairs, each of which had a single, similar-aged chick of its own (Postupalski & Holt, 1975).

As has been previously described, incubation in some raptors begins as soon as the first egg is laid. This egg, therefore, hatches in advance (2 or 3 days in the Sea Eagle) of any laid subsequently and this chick will be at an advantage over its younger and smaller sibling(s). The other(s) may suffer deprivation if food is in short supply or may even succumb to persistent attacks from an older and more precocious sibling(s). While this may result in what seems to be an unnecessary waste of young, the practice may be turned to advantage. Meyburg (1977, 1978) experimented with ways to prevent such nestling mortality in threatened species of eagles. The runt of a brood could be saved by removing it for hand-rearing or by placing it in the nest of another pair whose young are of similar age. As we have seen, such a technique was applied in Sweden. Meyburg, however, found that it was also possible to place the weakling chick in the nest of a totally different species, should there be none of its own species either available or suitable. When the youngster is almost ready to fly it can then be put back into its original nest to fledge as normal. If no suitable foster-parents are to hand, the chick can remain in human care but, to prevent imprinting on man, it is best to exchange the chick with a larger sibling every 10 days or so. Meyburg achieved notable successes with this technique amongst Lesser Spotted Eagles (*Aquila pomerina*) and the rare Spanish Imperial Eagle (*Aquila heliaca*).

In many cases a pair of Sea Eagles will rear two, or even three young quite easily but cases of serious deprivation or even death are known. Presumably because of the risk of desertion, few attempts have been made to employ Meyburg's techniques amongst wild Sea Eagles nor even to remove the first egg until the second is laid so that both young hatch simultaneously. Fiedler (1970) succeeded in doing just this with his captive pair in 1969 so that for the first time the pair managed to rear two young instead of one. Careful and prudent employment of such cross-fostering techniques or the adoption of captive-bred young into wild nests may possess great potential in rescuing threatened and rare species like the Sea Eagle from decline. Further decrease in numbers might thus be slowed, halted or even reversed until such time as the environment becomes less polluted. It is, however, quite a different matter to try to re-establish a species in an area from which it has disappeared.

Perhaps the most sophisticated of such attempts is the reintroduction of the Peregrine Falcon to its former haunts in eastern USA. At one time 250 pairs may have been distributed throughout the area but by the late 1970s none bred. A scheme to re-establish them is currently being operated by Cornell University (Cade & Temple, 1977). The most essential element in the whole operation is the highly successful captive-breeding facility from which derive the stock for release. A female Peregrine will normally produce a clutch of three or four eggs but under experimental conditions this can often be enlarged by removing single eggs or clutches as they are laid; as many as 16 eggs have been 'pulled' from one single female. The eggs are then incubated artificially and hatched in an incubator, the chicks being hand-reared or fostered back on to adults. Once the young are able to tear up prey for themselves – at about 4 weeks of age – they are removed to an artificial eyrie which has been constructed at the locality chosen for recolonisation. At first the young are confined by a barred partition and food is dropped from above through a chute to the eyrie below. Once the young can fly the door is removed so that they can fledge normally. Although free-flying they still return to the eyrie where food is being provided. A careful watch is maintained since the young do not benefit from adult protection against predators such as Great Horned Owls. After a further 2 or 3 weeks, the young falcons have taught themselves to hunt and kill and they gradually become less dependent upon the food provided. By 1982, over 400 captive-bred Peregrines had been 'hacked' into the wild, and recent breeding successes of several pairs have been noted (J. Barclay, personal communication).

In 1976, Cornell University was approached by the US Fish and Wildlife Services Endangered Species Unit to advise on a similar programme with Bald Eagles in New York, using some captive-bred stock but mostly young taken from wild nests. It is known that as many as 72 pairs of Bald Eagles may once have nested in New York State but by the early 1950s only a dozen pairs survived. By 1965, only one pair remained and since then (in 1973) has succeeded in rearing only one eaglet – an alarming consequence of the persistence of DDT residues in the local environment. Montezuma Wildlife Refuge in upper New York State, where Bald Eagles nested as late as 1959, was chosen as the release point: the lake nearby was proved to have minimum pesticide residues. The local power company donated and erected four telegraph poles, each 11 m high, upon which a suitable nest platform was constructed. On 27th June 1976, two captive-bred eaglets – one of each sex and aged about 9 weeks – were placed in the eyrie. They exercised so vigorously that a temporary cage had to be placed around them to prevent their falling out of the eyrie. Both eventually left the nest in mid-July but regularly returned to feed until they finally dispersed for the winter.

Over the next 3 years a further 21 eaglets were reared and released at Montezuma. Three are known to have died since. The original male from

1976 returned each year, but of the female there had been no sign until in 1980 both (recognised by their yellow wing tags) were found to have paired and to be in occupation of a territory by a small lake near Watertown, 80 miles to the north. They nested that year in a large Red Maple, hatched two eggs and successfully reared one young. This achievement, remarkable for birds of only 4 years of age, was repeated each subsequent year.

In the meantime, New York's only wild pair was still unable to breed successfully but in 1978 was rewarded for its persistence by being given two captive-bred eaglets to rear. The pair fostered two more the following year but in the spring of 1980 the old adult male was found shot. His widow quickly found a new mate – in one of the eaglets which had been 'hacked' from Montezuma in 1977. In 1981 and 1982, this pair, still unable to hatch eggs for itself, was able to continue in the role of foster-parents for captive-bred eaglets.

Thus, the 'hacking' programme has to date been able to maintain occupation of the original breeding territory and to add a new and successful breeding pair. Its ultimate aim, however, is to re-establish some 40 nesting pairs in the State and in order to achieve this it is estimated that 129 eaglets will have to be released. This they intend to implement over a period of 5 years. In 1981 they imported a total of 21 eaglets from eyries in Alaska, where a healthy Bald Eagle population still persists. These were accommodated in a 'condominium-style' tower at Oak Orchard – eight cages, 10 m up, overlooking Goose Pond (Figs. 59 and 60). All fledged successfully at the end of September and had dispersed 2 months later. Five have since died; two shot, one electrocuted and the others from unknown causes. In 1982 a further 21 were released, bringing the total liberated in New York State to 65 'hacked' eaglets and nine fostered (Nye, 1982 and personal communication).

Similar 'hacking' programmes, though not all on such a lavish scale, have now been initiated in several other states, such as Kentucky, Tennessee, Massachusetts and Missouri, under stringent guidelines laid down by the Fish and Wildlife Service.

Compared with North America, Europe seems slightly more reticent about reintroductions. Dwindling numbers of Eagle Owls (*Bubo bubo*) in Sweden are now being augmented using captive-bred stock (Broo, 1978) while a similar project is being envisaged for the Swedish Peregrine, which in 1975 was on the verge of extinction. An attempt is being made to re-establish the Raven in Holland (Timmerman, personal communication) and several Bearded Vultures (*Gyps fulvus*) from Afghanistan have been released in the Alps where the species disappeared early this century (Geroudet, 1977). In Britain, Great Bustards (*Otis tarda*), are being en-

Fig. 59. Rear view of Bald Eagle 'hacking' tower, New York State. The cages are open at the front only. A video camera, mounted on a pole in front, monitors the behaviour of the young birds, which are fed through hatches in the back wall.

Fig. 60. Bald Eaglets awaiting release from their cage, overlooking Goose Pond, New York State.

couraged (with little success as yet) to breed in captivity so that the young can be liberated on Salisbury Plain (Sharrock, 1976), while in recent years the Goshawk has managed to recolonise by means of both escaped falconers' birds and deliberate releases of hawks imported from northern Europe (Marquiss & Newton, 1982).

Like the Goshawk, the White-tailed Sea Eagle was exterminated in Britain by man. It is still much threatened elsewhere in Europe so that a successful reintroduction in a relatively unpolluted environment such as the coasts of Britain could make a significant contribution to the species' future survival on a global scale. The Osprey was able to stage a comeback on its own and found man's attitude to its presence was a more sympathetic one. The stage seemed set for the White-tailed Sea Eagle to return.

8 Reintroduction

Is mairg a théid do'n traigh 's na h-eòin fhéin 'ga tréigsinn.
It is a pity for the one who goes to the shore when the very birds are deserting it.

Gaelic proverb

Any translocation of a species from one area to another is not to be undertaken lightly. Three quite separate terms for the action may be identified. The first is INTRODUCTION which is defined in the World Wildlife Fund's *Manifesto on animal reintroductions* (1976) as 'the release of animals of a species into an area in which it has not occurred'. This is usually the most difficult to justify, even as a means of biological control of pest species, because the outcome may be totally unpredictable. There are too many instances of introduced species not performing as predicted, of increasing to pest proportions themselves, or ousting more desirable native species. The Rabbit is a prime example of an introduced species multiplying beyond all reasonable control and becoming, to farmers at any rate, a totally undesirable component in the ecosystem. There are many other examples. It was once a colonial fashion to release familiar European animals for completely spurious reasons. In Australia and New Zealand, for example, they now threaten many unique native animals and might have contributed to the extinction of others.

The second motive is RESTOCKING – 'the release of animals of a species into an area in which it is already present'. Prudently managed, this can be a useful conservation technique and one which, it might be argued, should be considered more legitimate in the face of the accelerating diminution of natural habitats. As an example we may cite the restocking of New York with Bald Eagles described earlier. Before embarking upon such projects, however, one must be aware of the reasons for the species' decline and whether suitable conditions still exist for its continued survival at this artificially increased level. It would seem but a short step to REINTRODUCTION – 'the release of animals of a species into an area in which it was indigenous until exterminated as a consequence of human activities'. It is into this final category that falls the White-tailed Sea Eagle in Britain, the topic of this book.

The WWF *Manifesto* (1976) goes on to identify essential criteria which must be fulfilled before any form of release into the wild of any animal is permissible. Briefly these are:

1. There should be an intensive study of the species and its environment past and present, upon which to base a firm objective basis for the reintroduction scheme.
2. It must not have a disruptive effect on the ecosystem in which it is carried out.
3. The catching, transport and release of the animals should be carried out legally, humanely and sympathetically in the first interests of the animals themselves.
4. A contingency plan should exist to discontinue the programme if the initial predictions are not satisfactorily fulfilled.
5. The local human population should be informed, on the whole sympathetic and not subject to serious economic consequences as a result.
6. Appropriate protective legislation should already exist.
7. The programme should be carried out objectively, scientifically and sensibly.

In addition they add two other criteria of particular relevance to reintroductions:

8. The animals used must be of the closest available stock.
9. The original causes of extinction have been largely removed, and the habitat requirements of the species are satisfied.

Thus it has been deemed necessary that a significant proportion of this book should be devoted to the White-tailed Sea Eagle, its biology, its worldwide distribution and its present status. After due consideration of various hypotheses to account for its decline, we may satisfy ourselves that man was a primary factor. The coasts of Northwest Britain and Ireland might yet support a viable population of the species had not various circumstances come about last century; an increasing human population suddenly became more coastal in its distribution; land-use changed to intensive sheep farming and game preserving, a process which encouraged mass disruption of the eagle's breeding and habitat, and introduced the means with which, through modern firearms and poisons, to effect its destruction. There are several factors inherent in the species' make-up which facilitated its extermination – its comparative approachability, the accessibility of its nests and its willingness to feed upon carrion so that it could more easily be poisoned. Many other raptor species simultaneously became reduced in numbers and some even became extinct in Britain, but a changing climate of attitudes together with protective legislation enforced since 1954 has encouraged the recovery of some (despite what hopefully proves to have been only a temporary setback from pesticides). The continued success of their recolonisation has been aided in no small way by an upsurge in public sympathy, generated in part by the publicity attached to the return of one popular species – the Osprey. Its return was a natural one, so active intervention to bring about the return of the Sea Eagle needs careful advocacy.

Naturally some sectors of the public will remain eternally unconvinced but it would seem that the fears of sheep farmers are unfounded, the Sea Eagle proving even less of a threat to lambs than is the Golden Eagle. Indeed, biologists would argue that all predators perform a useful sanitary function in our environment, and also among the prey populations themselves. Conservationists may express concern that such preferential treatment should be accorded only to rare and spectacular species at some expense to more mundane but equally deserving animals, but a bird like the Sea Eagle immediately attracts attention, awe and sympathy and by so doing can act as a worthy ambassador for the whole conservation movement, helping to promote its overall ideals. For years Norwegian ornithologists campaigned that some legal protection be awarded to eagles and it could perhaps be argued that an attempted reintroduction of Sea Eagles to Britain in 1968 might have helped convince the procrastinating officials. One final point should be stressed: biologists agree that a diverse environment is more likely to be a healthy and stable one. The continued presence of predators is a measure of diversity. In recent decades mankind was in acute danger of unwittingly polluting his own environment with no end of toxic substances. Being at the top of the food chain in that environment, and extremely sensitive to such chemicals, predatory birds provided an early warning that immediate and appropriate steps were necessary to avert global disaster. While a total solution to the problem is still a long way off, a more reasonable approach is being taken with some forethought for the consequences. Terrestrial systems are beginning to register signs of improvement but marine ones are slower to respond. We possess relatively few coastal predators convenient for research, birds being the most obvious and most popular. The Peregrine has already demonstrated its worth in this respect, while the Sea Eagle could also play a useful role. It may be in man's own future interests to encourage its return.

Few sympathetic conservationists would dispute the wisdom or desirability of the return of Sea Eagles to these shores, but some might question that man should take it upon himself to engineer it. Although the Sea Eagle became extinct around 1916, like the Osprey, it has as yet failed to recolonise of its own accord as the Osprey has done. The reason is likely to lie in a difference in the migratory habits. The Osprey undertakes an annual journey to winter in West Africa and each year birds were often reported in Britain, presumably *en route* back to Scandinavia. Some remained, mated and eventually bred. Two Swedish-ringed Ospreys have been known to breed in Scotland and the Scandinavian Osprey population is known to have been on the increase at this time. It would now seem to have established a reasonably firm foothold in Britain. In some parts of its range the Sea Eagle may forsake its breeding grounds to move south, especially in some Arctic regions where lakes may freeze in winter, but this is more of a dispersal than an annual migration and, on the whole, breeding adults are more sedentary than juveniles. Thus, if any Sea Eagles are to turn up in Britain they are more likely to be young birds with some years to go before they attain sexual maturity. In reality the frequency of sightings in recent times is very low indeed, too low it would seem for the species to re-establish itself. This is not to say that it might never have achieved successful recolonisation but, unlike the Osprey, its numbers are diminishing in Europe, so the time scale required might be more than this endangered species can afford. Thus, there are biological and conservationist justifications for a reintroduction, but I would also add a moral one – because it was *man* who brought about its extinction in the first place.

Just how he is to achieve this reintroduction requires careful thought. The most immediate results could be expected by implanting adult birds or already established pairs into vacant territories, but it is not easy to capture from the wild such adults and especially pairs, nor can it be assumed that the urge to breed or the pair bond will remain intact after the trauma of capture and translocation. Neither would there be any guarantee that such a pair would choose to remain and to breed in their new home. The use of captive adult stock for release is of questionable merit: the provenance of the birds to be released is often not known, and individuals which have spent much of their lives in zoos have been largely exempt from any natural selection pressures and will be totally unfamiliar with a wild existence.

A second approach would be to remove eggs or young from captive or wild pairs and to substitute them for the nest contents of suitable foster species. One species that immediately springs to mind is the Golden Eagle, which in some parts of Scotland is to be found nesting on the coast and, indeed, in some cases on the very nest ledges where Sea Eagles formerly bred. Although eggs and embryos are rather delicate, the problems involved in their transport are not insurmountable. Exchange of eggs has already been achieved with Bald Eagles in the USA (see Chapter 7). The main problem is possible imprinting of the young upon the wrong parents. Imprinting can be considered as a unique learning process whereby young animals identify with a mother object. We are all aware of Konrad Lorenz' goslings who adopted such objects as broomsticks, even the researcher himself, as their parents. The period of sensitivity of a young animal is brief, occurring just after birth or hatching, and once learnt is irreversible. In later life this may result in the total disruption of the bird's sexual activities and is therefore crucial in our argument. While our eagles might grow up and imprint upon the appropriate habitat, they may imprint upon the wrong species. This might be overcome by having two or more Sea Eagles hatching and growing up in one Golden Eagle's nest so that they could imprint upon one another rather than the parents, but Scottish Golden Eagles do not show much success in being able to rear two chicks. Regular supplementary feeding would therefore be required, incurring prolonged effort and regular disturbance at the nest.

Our precious eggs and eaglets would remain vulnerable to all the dangers of incubation and fledging, so that a successful outcome would not be assured. Some of these dangers might be overcome by substituting well-grown chicks into the nests of Golden Eagles, but some supplementary feeding would still be necessary and eyries are often remote and difficult of access. A simpler approach is to build an artificial eyrie on some suitable cliff ledge and for man himself to act as the foster-parent, much as is done with Cornell University Peregrines (see Chapter 7). Acquiring eaglets which are well-advanced in development, from wild nests or from captivity, keeping them together in pairs, and visiting them only briefly at feeding time would all help to reduce imprinting problems. Once the young are mature they can be allowed to fledge normally, yet still be provided with food near the release site until they can fend for themselves. This is a technique, called 'hacking', which has been employed for centuries in falconry where young hawks are released into the wild to acquire flying and hunting skills before being recaptured to fly at the fist. It is this very approach, modified to suit our particular requirements and conditions, which has been used to re-introduce the Sea Eagle to Britain.

Reintroduction 157

We recall that the WWF *Manifesto* (1976) had stipulated that the source population should be as close as possible to the original extinct stock; and, of course, it must not itself be endangered by removal of young. The only area to fulfill both these requirements is the north coast of Norway. We have already seen how it retains a healthy stock of Sea Eagles and how their habitat and habits are very similar to those once prevailing in Scotland. An additional advantage is that Norwegian Sea Eagles seem least inclined of any in Europe to disperse from their natal area; thus, once released in Britain the young are unlikely to leave again. Nonetheless, Norway is conveniently close to facilitate easy shipment. Most important of all, the authorities there have responded sympathetically and generously to our approaches. There remains the possibility of augmenting this source from captive-bred stock but no-one yet breeds Sea Eagles in Britain and to initiate such a facility would be both time-consuming and costly. Indeed, the number of captive pairs in Britain at this time is very low. Nonetheless we cultivate contacts with institutions both at home and abroad who may at some time in the future be able to co-operate in this respect.

Since no equivalent reintroduction scheme has been attempted before, we have no idea how many Sea Eagles we may need to release to achieve a self-sustaining population. Perhaps we could find a useful parallel in the Icelandic situation where there is little, if any, input of immigrants to augment local breeding output. Despite a minor degree of persecution still, the species has maintained a precarious foothold in the country, with as few as 20 pairs, sometimes fewer. It may not be coincidence that 20–30 pairs is the current level attained by the Osprey in Scotland, a state of affairs which is giving conservationists some cause for self-congratulation. (Scottish Ospreys do enjoy additional benefits of occasional immigrants arriving from the continent.)

If we assume a survival rate among our Sea Eagles of around 50%, to establish a population of 20 breeding pairs we need to release a minimum of 80 eagles – possibly 100 to allow a margin for safety. We shall see presently that the survival of the eagles being released on Rum may be in excess of 50% and that our annual injection of six eaglets has permitted our pioneer population to increase. Thus we could foresee that the time to cease further importations is when our first pairs are breeding and themselves adding six locally-bred recruits each year. In the most favourable habitat in Norway it has been shown that some 60% of breeding pairs prove successful, each rearing about 1.6 young per year. Since our stock is to be of Norwegian provenance, from an environment in many respects identical to Scotland, we may expect them to aspire to an optimum breeding output. Thus, to achieve our target of six young we would require at least seven pairs to be breeding in the wild. Initially, however, our pioneers will be young and inexperienced. Furthermore, the smaller our population, the more vulnerable it will remain: one irresponsible act of poisoning, for example, could prove a major setback. Thus we should err in favour of the eagles, and obviously, the more we are prepared to do so, the greater our likelihood of ultimate success.

Before describing in detail this latest reintroduction attempt on Rum, it is first necessary to mention two others which are known to have been made in the recent past. The first of these, and possibly the first attempt ever made in Britain, was undertaken by Mr Pat Sandeman in July 1959. He had obtained an adult and two young birds from Norway, where someone had captured them to claim bounty. With the permission of the estate owners, the three Sea Eagles were tethered at a spot in Glen Etive in Argyll. A red deer carcass was left within their reach and a fortnight later all three were released. Unfortunately, the adult was rather tame and soon appeared by the roadside where it posed for the cameras of passing tourists! About a month later it was caught red-handed attacking the chickens of an Appin farmer who, thinking it to be a Golden Eagle, despatched it to London Zoo. Properly identified it was eventually returned north to take up residence in the zoo at Edinburgh. The two juveniles, being much wilder, quickly learned to fend for themselves and remained in the neighbourhood for several months. In January 1960, nearly 6 months after its release, one was killed in a fox trap at Otter Ferry, some 50 miles to the south. Of the other, nothing more was heard although a Sea Eagle was said to have been spotted about this time at the Mull of Kintyre. Perhaps it still roams the seacliffs of the west awaiting a mate (Sandeman, 1965).

A second reintroduction attempt took place in 1968 (Dennis, 1968, 1969) and was financed by the Royal Society for the Protection of Birds under the guidance of the late George Waterston, at that time the Society's Scottish Director. The venue was Fair Isle in Shetland where Sea Eagles last nested around 1840. The island was considered to be suitably remote, with an abundance of prey species and, no less important, with a local community who were amenable to the whole scheme. Fair Isle is owned by the National Trust for Scotland and the then warden of Fair Isle bird observatory, Roy Dennis (together with his assistant, Tony Mainwood) were to provide the necessary expertise and manpower. Four eaglets were obtained from eyries in Norway by Dr Johan Willgohs in mid-June 1968; a male and two females were flown to Fair Isle on 24th June, and a second male about 2 weeks later. Two temporary cages had been erected on a hillside near seacliffs known traditionally as 'Erne's Brae'. Each cage consisted of two compartments, each 4 m square, and provided with a small box-shelter and a perch. The birds were thus kept as pairs, separate, but within sight of a sibling.

One female, named Torvaldine, precocious from the outset, fed boldly from the hand; the other, Ingrid, preferred to take food placed on the ground in front of her. The male, Jesper, was shy and cowered in one corner of his cage but after his first night was found to have fed himself. All three were about 7 or 8 weeks old on receipt and had reached maximum weight although had still to complete feather development. As the flight feathers grew, the birds began to exercise and the first flight within the confines of the cage was made on 20th July. The latecomer, Johan, was younger than the rest but soon caught up with them. About 750 g of meat or fish was provided to each bird every day but not all of this was eaten. Later the eagles became more shy and restless and, so that they would not become tame, visits with food were made only every 2 or 3 days.

Prior to its release, each bird was caught for weighing and measuring and each was given a metal numbered ring on one leg, with an individual colour ring on the other. Ingrid was the first to be set free, the wire mesh on one wall being loosened in readiness. The next day, 16th September, the wire was pulled back, using a long cord from a hide constructed some distance away, thus permitting the bird to leave undisturbed and in her own time. With a seeming flair for anticlimax, Ingrid decided to walk to her freedom and spent the next 30 minutes scaling a small hillock nearby. It was only when a Raven landed briefly beside her that she took to the air for her maiden flight. Her second flight was more prolonged although she had to endure a mobbing from a passing Peregrine. Within a couple of days Ingrid had become more adventurous but still suffered persistent attacks from other birds, mainly crows and gulls. Although food had been dumped at a conspicuous site for her use, she preferred to search the beaches for carrion. On 21st September, after 5 days of freedom, she was observed eating a freshly dead Oystercatcher (*Haematopus ostralegus*) but it is not certain whether she had managed to kill this herself. On 2nd October she was joined by Jesper, and then by Johan 2 days later. All three returned regularly to the empty cage to feed on rabbits and dead birds which were being put there as a food dump. When Torvaldine was released on 20th October, a new food dump had to be located on the cliff-top. During this week, in a spell of bad weather, Johan disappeared, probably wandering out to sea; he was not seen again.

The three remaining eagles were often seen displaying and calling together during the ensuing winter. The male would swoop down on a female, who rolled over with spread wings to present talons. Occasionally a pair would grapple and tumble out of the sky a short distance before disengaging. They continued to utilise the food dumps but were also beginning to find food for themselves. An eagle pellet was found which contained fish bones, while the birds were seen feeding on dead birds or an occasional seal carcass on the shore. From 6th March, Ingrid began to lead a solitary existence while Torvaldine and Jesper continued to frequent the food dumps; he preferred to feed on the remains of whatever Torvaldine had carried off for herself. Ingrid was last seen on 12th April 1969, 7 months after her release. She was apparently in good condition and probably left the island during a spell of clear weather when Shetland (to the north) and Orkney (to the south) were visible on the horizon.

The first moulted feathers were found on 9th April at roost sites on the cliffs and the wing and tail moult became obvious by the end of the month. At this time the eagles sought refuge from persistent mobbings on the cliffs and spent less time in the air. The lambing season was then in full swing and although once or twice the eagles hovered curiously over a ewe with her lamb, they made no attempt to attack. Nor were they seen to chase Rabbits, although these displayed no fear of eagles overhead. Torvaldine and Jesper must have been, by then, largely self-sufficient and carrion was obviously an important constituent of their diet. The first indication of an actual kill was on 8th May when one of the eagles was seen to catch a Fulmar in flight. It released it almost at once but later caught another which was held for longer before it was let go. On 20th May, Torvaldine was seen carrying a glass bottle in her talons, which she had presumably snatched from the surface of the sea in mistake for a fish. Thereafter the remains of freshly killed Fulmars began to be found on the cliff-tops.

In clear weather the eagles often soared to 700 m or more and even ventured 4 or 5 km out to sea. Eventually, Torvaldine left the island during the second week of June, about 8 months after her release. Only Jesper now remained and he continued to catch Fulmars despite his heavy wing moult, also two young Shags taken from a nest. He remained hidden amongst the cliffs for the next weeks but the remains of a young Fulmar, a Shag and a gull were found on the shore. On 19th August he was flushed out of a sea cave but, unable to fly properly, fell into the water; he drifted ashore and was caught. Jesper was fat and well-fed but his plumage was smeared in Fulmar oil. It is thought that he had been approaching young Fulmars on their nest ledges only to be spat at repeatedly before he finally despatched his victim. He was set free again but probably died soon afterwards, being last seen on 28th August. He had survived 10 months in the wild and died in rather unexpected circumstances. Although several other predators have since been known to have become contaminated with Fulmar oil (Dennis, 1970) this must have been an unusual occurrence. Sea Eagles can, and do, regularly catch Fulmars, either from the surface of the sea or in flight, apparently without any risk of being spat at. It proved

Jesper's undoing that he should opt for an easier alternative by walking up to Fulmar chicks looking deceptively helpless on their nest ledges.

No other eagles were released on Fair Isle. Yet the project cannot be dismissed as unsuccessful. It is possible that one or more of the three have survived, two having demonstrated that they were capable of looking after themselves. Since 1968 there have been several vague and unconfirmed reports of Sea Eagles, some of the more recent ones apparently involving adult birds and therefore possibly of Fair Isle origin. However, the release programme involved so few birds and in one year only that a successful outcome was unlikely. Nonetheless, it proved the feasibility of a method which could be employed in any further reintroduction attempts.

It was not until 1975 that a fresh attempt to re-establish Sea Eagles in Britain was begun. Several venues were considered. Although Fair Isle had seemed ideal it proved to have several disadvantages: it lay in the open sea some 35 km from either Orkney and Shetland and its 810 ha rose to only 217 m. It thus would present a tiny and inconspicuous outline on the horizon to any Sea Eagle which might stray too far afield.

Since the new project was being managed by the Nature Conservancy Council, a National Nature Reserve seemed appropriate. St. Kilda was one possibility but being remote, like Fair Isle, Sea Eagles may not find it easy to colonise elsewhere in the Hebrides: it was well over a century since the species had last bred there and – last but not least – the many thousands of Fulmars which nest on St. Kilda could have presented a problem to young and inexperienced eagles.

The island of Rum (Figs. 61 and 62) in the inner Hebrides seemed suitable in all respects. It lies in the heart of the Sea Eagle's former range in the West of Scotland; if the birds wandered they were still within suitable habitat. The last known pair had bred on Skye in 1916 – only 13 km away – while a pair had last nested on Rum itself barely a decade earlier, in 1907. The island is large, extending to 10 600 ha, and lies only 24 km from the mainland. Its impressive range of hills culminates in several peaks over 800 m and presents a silhouette which is famous and familiar throughout the Hebrides. Rum's extensive rocky coastline supports colonies of Eider, Shag, auks and gulls – all potential prey for Sea Eagles. Fulmars on the other hand, are relatively scarce, the 400 or so pairs having a localised distribution mainly on the southeast coast (Love, 1977, 1980a). The hills of Rum harbour a unique colony of Manx Shearwaters (*Puffinus puffinus*), numbering in excess of a 100 000 pairs (Wormell, 1976). Despite their visiting nest burrows only at night, some Shearwaters fall prey to Golden Eagles and

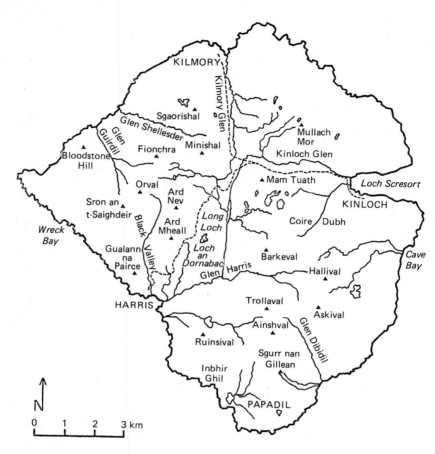

Fig. 61. Map of Rum.

could similarly be available to Sea Eagles. A source of carrion on Rum is provided by the population of 1500 Red Deer (*Cervus elaphus*) and 200 Feral Goats, while the fishing grounds around the island are as productive as anywhere in the Hebrides. Both otters and gulls are abundant around the shore and from them the Sea Eagles could pirate fish. There are no sheep, rabbits nor hares on the island but otherwise the island seemed eminently suitable for the reintroduction of Sea Eagles. The Nature Conservancy Council (NCC) maintains a community of eight families on Rum and, since it became a reserve in 1957, public access is restricted.

In 1826–28 Rum's entire native population of some 350 souls were evicted and transported to Canada; the ruins of their houses and cultivations are still to be seen around the coast. The island then became a sheep walk but by 1845 the emphasis had turned to Red Deer (which had by

Fig. 62. Isle of Rum from the east.

that time become extinct): animals were reintroduced and encouraged to multiply. At this time there were at least five pairs of eagles breeding on the island. Harvie-Brown & Macpherson (1904) indicated that these were all Golden Eagles, with some additional pairs of Sea Eagles, but I would consider it unlikely that the island could support much more than five pairs of Sea *and* Golden Eagles. Some persecution occurred but several pairs persisted until 1890, but, with the proprietor's encouragement, the Sea Eagles and possibly the Goldens too were exterminated in the next two decades. The latter have subsequently recolonised under their own volition and have since increased to the existing level – interestingly, five pairs.

A scheme for the reintroduction of Sea Eagles was devised by Ian Newton of the Institute of Terrestrial Ecology and instigated by Martin Ball, Deputy Regional Officer of NCC, together with Dr J. Morton Boyd, its Scottish Director, Roy Dennis of the RSPB also being approached for specialist advice. I was recruited in May 1975 to manage the project on Rum, together with the island's then Chief Warden, Peter Corkhill.

The approach to be employed was to be similar to that on Fair Isle in 1968 and with which Roy Dennis had first-hand experience. We were fortunate also to enlist the willing assistance of Dr Johan Willgohs and through him the Norwegian authorities generously granted permission for the export of our first eaglets. Transport to Fair Isle had been by expensive private charter. Rum possessed no airstrip but before any undue complications arose the Royal Air Force at Kinloss responded favourably to our tentative approaches. It transpired that they made several operational sorties from Kinloss on the Moray Firth to northern Norway and a Nimrod of 120 Squadron was enlisted to come to our aid. Johan Willgohs was to be engaged in a survey of Sea Eagles in the Bodø area towards the end of June and agreed to collect for us four eaglets. A rendezvous on 22nd June was arranged with the RAF at the conveniently situated Norwegian Air Force base at Bodø. Final preparations could now proceed.

I took up residence on Rum in early June and supervised the siting and construction of suitable cages. Their ultimate situation was determined by Government quarantine regulations which stipulated that the eagles were kept at least 8 km away from domestic poultry. A locality was selected which was sufficiently remote from sightseers but accessible (along 10 km of what is optimistically referred to on the island as a road). The area overlooking the sea offered ample opportunity for the eagles, upon release, to scavenge or hunt along the shore. The cages were to be of a design similar to that used on Fair Isle, and constructed of wire mesh stretched over a timber framework. Two compartments, each 4 m square and 2 m high, were large enough to permit the occupant to exercise freely but sufficiently restrictive to prevent the eagle's damaging itself against the mesh walls. The birds were to be kept as pairs and my visits to feed them were to be brief and infrequent to prevent their imprinting on anything other than themselves. Each bird was provided with an open-fronted shelter 1 m square, and two log perches. Two cages were constructed in 1975 and two more in 1976, incorporating minor improvements.

In the meantime Martin Ball had been in Norway assisting Johan in his eagle survey along the coast north of Trondheim: in mid-June I flew out to replace Martin at Bodø. This bustling fishing port, with a population of over 30 000, is the capital of the Nordland commune. Lying just north of the Arctic Circle, it is a favourite resort for tourists wishing to witness the midsummer sun but it also lies in the heart of one of the most dense Sea Eagle populations in Norway. The eagles' main haunts were the myriads of offshore islands, large and small, although they also nested on the mainland coast and fjords. Several of the main island groups supported small communities of fishermen and farmers, some of whom supplemented their incomes by harvesting Eider down, collecting snowberries or even hunting otters for skins. The local people all shared a lively interest in wildlife and several proved invaluable contacts for Johan in his quest for eagles.

The Sea Eagles were not too difficult to locate. At each of two communal roosts – in birch trees that clothed a steep slope on two of the larger islands – we flushed 15 to 20 immatures and sub-adults. Beneath the trees where the Eagles had perched the ground was littered with droppings, down and moulted feathers. In the thick bracken it was difficult to find many pellets or prey remains; nest sites were more productive in this respect. In the course of my 2-week sojourn we identified 52 prey items: 60% were fish (mostly Catfish, but also Lumpsucker, Flounder, Angler-fish and gadids. The remainder were birds, with Eider comprising up to 75% of the sample but also some Shags, auks and the occasional gull or crow.

The eyries we visited were mostly located on broad cliff ledges, some on more steep slopes and one was even lying like a survey trig point on a high spot of a flat, remote skerry. Two eyries had been constucted in trees, one having since collapsed under the weight of the nest structure. All of these nests were accessible except for one on a vertical cliff face by the edge of a mainland fjord. Here, Johan explained, seabirds were less abundant and fish, especially deep-water species such as Cod, featured in the diet of the eagles. These 'inland' pairs also took a few small mammals and carrion in the winter months. Four of the eyries contained a single chick and another four contained two. We were told of an eyrie in the vicinity which had three young, while the collapsed tree we had visited earlier had contained a rare clutch of four eggs. As we ringed each chick the parent eagles would circle anxiously above the nest yelping loudly. It was easier to distinguish the male of the pair by his smaller build and higher-pitched voice. Nearly all the eaglets were about 7 or 8 weeks old. The chicks of course hatch several days apart and at one eyrie in particular we were impressed by one precocious individual leaning back unsteadily on his huge yellow feet, hissing defiantly at our intrusion while his older sibling cowered on the floor of the nest like a frightened fawn.

At the end of the first week, reaching a latitude of 68° North, we were forced by a storm to retrace our course for temporary shelter in Bodø. Somewhat bemused at our sudden return to civilisation we soon gained courage to strike out into the elements and made for an attractive offshore archipelago, rich in seabirds and containing one of the few Cormorant (*Phalacrocorax carbo*) colonies in northern Norway. These islands had become a nature reserve in 1967 and one of the score or so local inhabitants had been appointed warden. One eyrie he took us to, of the five occupied that year, had been robbed a few days before: powerboats from nearby towns sometimes raided eyries to sell eaglets to foreign falconers, but on the whole, human interference and persecution did not appear to be a serious problem in Norway.

Reluctantly we took our leave of these islands, which reminded me so much of the Outer Hebrides, but the persistent bad weather introduced some doubt as to whether we could meet our deadline with the RAF. In desperation, armed with the necessary licences and permission from the local landowners, we secured four chicks from two of the nearest convenient eyries. The eaglets lay quietly in their roomy cardboard boxes during the cruise back to Bodø where they readily accepted from me their first meal. Johan took me to the airbase, where I was offered accommodation, before he continued on his way. Overnight the eaglets left an indelible impression on the walls of the shower room attached to my bedroom and next day continued on their way to Scotland. For some reason of security I was not permitted to accompany them, much to the apprehension of the Nimrod's crew. My lengthy detour, by civil airline from the Arctic to Inverness *via* Oslo and London, and thence overland and by boat to Rum, took several days, whereas the eaglets reached their final destination within only 11 hours of leaving Bodø.

This same procedure has been followed each year subsequently, except that the collecting operation has been undertaken by Captain Harald Misund, a keen ornithologist whose knowledge of the Bodø area and its Sea Eagles is unsurpassed. Due to his tireless efforts prior to our arrival each June, the eaglets are almost always ready and waiting for us. In almost every instance they have been single chicks from broods of two so that the parent eagles have been left with a chick to rear to fledging. In one instance a runt from a brood of three was taken: it would certainly have died but once in captivity thrived and was released into the wild again, but in Scotland not Norway. The eaglets derive from different regions each year and only in three instances have we ever deprived a pair of more than one chick. Thus the stock which we are introducing to Scotland is highly variable genetically. The eaglets are given, free of charge, by the Norwegian authorities and the local landowners, and Harald receives no remuneration other than his minimal expenses, which are covered each year by WWF Norway whose Projekt Havørn in 1978 pledged their continued support by donating free of expenses two eaglets: in 1980 Harald himself presented us with two more. All other expenses (other than the RAF flight from Bodø of course) are borne by NCC, aided in 1980 by the RSPB, WWF and the SWT.

In June 1975 we received one male and three female eaglets. An early disappointment, however, was the death of the only male (named Odin) just as he attained fledging. He failed to respond to treatment and all that could be diagnosed from a *post mortem* was kidney failure – a condition sometimes induced by stress, perhaps from the original illness itself. The three females all fledged successfully.

Most of the eaglets arrive on Rum at an age of about 8 weeks (Fig. 63). A

Reintroduction

Fig. 63. Seven-week-old eaglets.

few younger individuals demand hand-feeding and one or two of the most reluctant even force-feeding initially. As soon as the eaglets can use their feet for tearing up prey (Fig. 64) and moving around, they threaten any human contact (Fig. 65). Once they can fly, at 10 or 11 weeks, they become distinctly nervous at my approach. Often males tend to be more highly strung than the females and on occasions feathers can be abraded against the wire: a male, Cathal, (Fig. 66) also broke off the tip of his beak in the mesh (it soon regrew). So in 1976 when we received 10 eaglets we devised a tethering system. At that time we intended retaining four juveniles ultimately to breed from them in captivity. Adapting a traditional falconry technique we attached to each tarsus leather jesses (Fig. 67) which met at a metal swivel: a 1 m length of braided nylon rope served as a leash, terminating in a metal ring. Threaded through a 6 m metal cable, stretched taut between two short posts, this permitted the eagle to fly from a tree stump perch at one end (just within reach) to a tent-like wooden shelter at the other whose roof acted as a high perch.

The females Colla (Fig. 68) and Sula, a male, Ronan, and Beccan (thought at that time to have been a male also) were retained in long-term captivity. Colla and Beccan both escaped temporarily but Beccan tried again,

168 The return of the Sea Eagle

Fig. 64. Eaglets about 8 weeks old tearing up prey, Rum, June 1980. (Photo: J.A. Love)

Fig. 65. Juvenile in full threat.

Fig. 66. First-year male, Cathal.

successfully, necessitating replacement by Cathal. These mishaps led us to improve upon the strength of the metal apparatus used while the leather 'furniture' – improved over several centuries by falconers – proved totally adequate. In September 1978, Colla took ill suddenly and died a few days later. Not only were we unable to summon veterinary help in time but we were prevented by transport delays from despatching the corpse in time for an adequate pathological examination. We suspect that Colla had died of a bacterial or viral infection perhaps transmitted by the rats or crows which frequented the tethered site for scraps.

In 1978 we imported eight eaglets and one of these, a female, Shona, seemed to suffer from a mineral deficiency; after a vigorous bout of exercise, not long after her arrival, she fractured both legs. A visiting doctor temporarily set her brittle bones until the eagle reached a vet on the mainland and eventually the Veterinary College in Glasgow where the fractures were pinned surgically. Two local ornithologists, George Watt and Mrs Carol Scott, kindly agreed to let Shona convalesce in their back garden (in Eaglesham, appropriately!). Under their expert care, Shona recovered sufficiently to return to Rum in October, but by then she had lost all fear of humans and still walked awkwardly. She had strong callouses around the fractures but once on tethers complications arose at her wing joints.

Fig. 67. The author fixing jesses to a hooded juvenile about 10 weeks old. Rum. July 1980. (Photo: J.A. Love)

Incapable of normal flight thereafter she caught a chill, perhaps pneumonia, in September 1979 and died. This time a prompt *post mortem* at the North of Scotland College of Agriculture in Inverness revealed a severely congested bile duct and gallstones (which could have contributed to, if not caused, her death).

The sad losses of Odin, Colla and Shona highlighted the remoteness of Rum from speedy and specialised veterinary advice, deficiencies which

Fig. 68. Sub-adult White-tailed Sea Eagle, Colla.

could be crucial should we set up a captive-breeding facility on the island. At that time too, financial restraints were imposed on NCC as a government-aided department and we were conscious that Harald wished all the birds to be liberated. He assured us of further young in future years so we decided to abandon captive-breeding altogether. In the spring of 1979 Cathal, together with the original captive pair from 1976 – Sula and Ronan – were set free.

All six eaglets imported in 1979 were released that autumn, eight more in 1980, another five in 1981 and 10 in 1982.

While in captivity the eagles were provided with as natural a diet as possible. All their food was obtained fresh and locally, although some was then deep-frozen for later use. During the summer months it was possible to catch an abundance of Mackerel offshore, and we sometimes begged other fish from local boats or from the fish market in Mallaig; thus Whiting, Haddock, some Herring, Dogfish and Squid could be added to the eagles' menu. By their ability to digest most of the smaller fish bones, the eagles derived a source of minerals, calcium being especially important for growth of their bones and feathers. Fish also contained sufficient moisture to satisfy the eagles' water requirements but the long-term captives had access to water baths in any case. In a climate precipitating over 250 cm of rain *per annum* this seemed an unnecessary luxury! Apart from occasional small pellets of dried grass, the eaglets regurgitated little waste on their diet of fish. We gave them whole bird carcasses as a source of roughage. These were mostly gulls and crows shot with a rifle, to prevent the possibility of the eagles accumulating quantities of potentially poisonous lead shot. Towards the end of the breeding season, young gulls were available from convenient local colonies. Many of their smaller bones could be digested by the eagles but the larger ones, the feet, beak and feathers were usually regurgitated in plump, wet pellets; the keel bone and wings were usually picked clean and discarded. In early August scraps of venison, viscera and heads of shot deer became available from the stalking activities on the island, and this was supplemented by goats shot along the coast. Waste food was removed daily in the interests of hygiene and rats cleaned up remaining fragments. I never evidenced the eagles attempting to catch or kill these rats but on at least one occasion the tethered male Ronan succeeded in capturing and killing an over-confident Hooded Crow which had boldly strayed within reach, in its search for scraps. At times crows and Ravens would prove a nuisance by pulling food beyond the reach of the tethered eagles. At times I would discourage the corvids with my rifle but also made sure that the tethered eagles had generous rations of food.

The amount of food consumed by the eagles varied from day to day – from almost nothing (usually following a hefty meal the previous day) to over 1.4 kg. On average I found that a captive eagle would consume about 10% of its own body weight per day, about 450–600 g wet weight (see Chapter 5 and Love, 1979). Two or 3 days' supply could be stored in the crop but most of this would pass through the digestive system within a few hours. Once having fed, the captive eagles tended to sit lethargically, looking around or preening.

Thus, over the eight seasons since 1975, we have handled a total of 55 eaglets. Although no attempt was made to sex them in the nest, it has transpired that we have received 30 females and 25 males. Three eaglets (5% of our imports) have died in captivity – two females and a male. One would expect a proportion of 'non-viable' birds which would be eliminated from a wild population during fledging or immediately on leaving the nest. In captive conditions such poor individuals may survive longer but would still be likely to succumb on release. So we must not be too disappointed should some losses occur at this stage of the operation.

Fig. 69. Juvenile, tethered and hooded.

Immediately prior to its being given its freedom, an eagle is caught and hooded (Fig. 69) so that we may measure and mark it. Measurements taken of bill length, from the cere to the hook tip; and of the bill depth, from the cere to a point perpendicularly below the lower mandible as it is held closed (Fig. 70). In 1977, additional statistics were gathered on the thickness of the tarsus (at its narrowest point midway along its length) (Fig. 71). Adequate samples of wing, tail and tarsus length can be found in the literature (Willgohs, 1961; Cramp & Simmons, 1980) and in any case abrasion can confound the measurement of feathers. Finally the bird was wrapped in a canvas sling for weighing, allowance being made should there be any food in its crop (see Helander, 1981c). In all but three instances the crop was deemed empty. Together these five statistics (bill length and depth, tarsus width and thickness, and body weight) were sufficient to sex the eagles (see Chapter 2) and in most cases were found to confirm my own, purely subjective, conclusions based upon the bird's appearance, voice and behaviour. Several birds fell within a zone of overlap between males and females, but I discovered that a combination of bill depth and tarsus thickness gave the most reliable result (Fig. 14). Only two birds do I consider in retrospect to have sexed wrongly – both early in the project when I was less experienced and had less data to hand. At first I considered Beccan to be a male, which could have proved an unfortunate setback had we tried to give 'him' a mate in captivity!

Each bird was provided with a British Trust for Ornithology (BTO) metal ring bearing a serial number and address, so that if found the origins of the bird could be traced. On its other leg, I provided each eagle with two coloured plastic rings of Darvic sealed with chemical fixative. These facilitate individual recognition in the field. The colour of the lower ring denotes the year of fledging – black for 1975; white for 1977; pale green for 1978; red for 1979; dark green for 1980; orange for 1981; dark blue for 1982. The top colour made the combination unique to any one individual. In 1976 I used a variety of combinations but placed them all on the right leg; in all other years the colour rings were and will be placed on the left leg, with the BTO ring on the right.

Reintroduction

Fig. 70. The author and NCC warden Alex Scott measuring the bill depth of a hooded juvenile Sea Eagle just prior to its release, Rum. (Photo: J.A. Love)

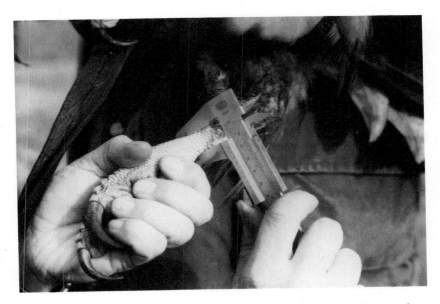

Fig. 71. Measurement of tarsus thickness; the long, sharp claws and spiky sole of the eagle's foot are clearly shown, Rum. (Photo: J.A. Love)

9 Release

I hear the eagle bird
With his great feathers spread,
Pulling the blanket back from the east.
How swiftly he flies,
Bearing the sun to the morning.

Iroquois poem

The instant I whip the leather hood from its head the eagle blinks and spins round to threaten my movement behind. One eye stares at me suspiciously while the other slowly begins to register unfamiliar surroundings beyond. At last the eagle opens its wings and with a loud 'swish' is borne away with the wind. At this moment in each release I am conscious of the eagle's seeming to hesitate, expecting itself to be tugged back to earth by its jesses . . . but no, this time it is free. It beats its ponderous wings to gain height, seeking control of the wind with its straining pinions: its legs dangle awkwardly beneath for never before has it had occasion to tuck them up (Fig. 72). On this its first sustained flight it spontaneously discovers the rudimentary principles of flying. When its undercarriage is finally stowed the bird is more stable against the sky and negotiates a huge semi-circle. A passing Raven stoops idly from above but the eagle can counter defiantly by tilting on one wing to flaunt its outstretched talons. With a few more flaps the eagle circles again, gradually loses height, wobbles uncertainly and tumbles into the long grass. It has chosen a gentle, thickly vegetated slope free from rocks and other impediments, for this is its first landing at high speed. Hastily the grounded bird recovers its dignity and, sitting upright to

Fig. 72. Release.

survey the landscape, it seems to register distinct bewilderment at this strange new turn of events. Thirty minutes later it begins to shuffle laboriously through the long grass to reach an inviting knoll nearby, where it may then remain for several hours.

A Hooded Crow appears beside the eagle, expectant of a tasty scrap, but perceiving no food it boldly delivers a hasty tug of annoyance at the huge bird's tail before fleeing. The eagle reacts sluggishly and, swivelling round, it catches a gust of wind under its half-opened wings and is lifted off on its next brief flight. The sorties become longer and more adventurous as the eagle makes its way along the hillside. Finally it is dusk and the bird displays little initiative in its choice of roost, preferring simply to move into the lee of a large boulder where it spends the night.

At times, however, a release can prove a dreadful anticlimax. Embarrassingly, one eagle chose to walk out through the open door of its cage and set off, with a hunched but determined shuffle, to scale a prominent knoll nearby. If possible, I now like to liberate eagles when birds released in previous years are in the vicinity. This can often stimulate more excitement so that on a maiden flight a youngster can be induced to execute quite complex aerial manoeuvres. But if sometimes lacking in drama, the moment of release is thrilling, certainly for me, and I can't help but feel, for the bird too. Such magnificent creatures were not brought into this world to be held captive: it is especially rewarding to accord them their freedom, knowing that one day some may choose to breed on our long forsaken shores.

Although the eagles may have been able to fly within a few weeks of arriving in this country, they are retained on Rum for a further 6 or 8 weeks. Part of this period is the mandatory month of quarantine, and during the rest we might expect them to be most dependent both physically and psychologically upon their parents. Only eight have been released during the spring months – four of them being the long-term captives Ronan, Cathal, Beccan and Sula.

Food is always available to them in the vicinity of the cages or tethered site (Fig. 73). The presence of eagles which are still captive usually attracts the youngsters back (Fig. 74). Their social nature is demonstrated in that these birds prefer to steal food from the tethered birds rather than to feed on their own at a deer carcass lying only a short distance away. Once a routine is established the 'decoys' themselves are liberated. The juveniles usually team up together and often, too, with older birds released in previous years, benefitting from their experience. It is reassuring to see how tolerant the old birds are, some even willing to yield up food to the juveniles, so that their survival prospects are much enhanced – an important by-product of the project's being sustained over several years.

The very first Sea Eagle released on Rum (see Table 14) had no such advantage. I had called it Loki, suspecting it to be a male, but once fledged it was obviously a female. On 27th September 1975 I opened up the door of Loki's cage but frustratingly found her reluctant to emerge. A fresh attempt was made the next morning by removing the wire from one wall of her cage. We watched from afar as she leapt at where the wire had been and literally fell out of her cage to freedom. Frantically flapping her wings she became airborne and found herself carried across the glen, with her legs dangling absent-mindedly, before crash-landing on the far slope. Not quite the drama we had anticipated for our first release. After several minutes she regurgitated a fat pellet. I remembered how she had gorged on two Red Grouse (*Lagopus lagopus*) the day before, which would explain her lack of co-operation that day. Before dusk she made only one or two brief exploratory flights but the next morning I spotted her soaring competently at the top of the glen. Twice she was attacked by a juvenile Golden Eagle but skilfully avoided it in the air. Ten minutes later two adults joined in the fray but soon lost interest and moved on. Within two days I found that a Greater Black-backed Gull carcass had been carried from the food dump, and on the fourth day I watched her eat some deer liver at the food dump. Thereafter she began to wander further afield but would regularly return to sit on top of the cage which yet contained the female Karla. By mid-October Loki had

Fig. 73. Sub-adult and juvenile Sea Eagles at food dump.

Fig. 74. Three tethered eagles awaiting release with a juvenile (far left), already liberated, returned to the food dump to feed. The fence prevents undue disturbance from deer and cattle, Rum. (Photo: J.A. Love)

Table 14. *Sea Eagles released on the Isle of Rum: 1975–82*

Year of import	Males			Females		
	Name	Weight (kg)	Release date	Name	Weight (kg)	Release date
1975	(Odin	4.2 (dead)	died)	Loki	5.9	27.9.75
				Freya	6.9	1.11.75
				Karla	6.4	24.10.75
1976	Fionn	6.5	23.11.76	Iona	6.25	3.5.77
	Ailsa	5.75	23.11.76	Isla	6.0	15.4.77
	Cuillin	5.4	24.1.77	Beccan	6.5	4.5.77
	Brendan	5.2	28.2.77	(Colla	6.5	died)
	Ronan	5.1 (x)	24.4.79	Sula	6.0 (x)	6.3.79
1977	Mungo	5.1 (x)	18.10.77	Vaila	5.8 (x)	18.10.77
	Cathal	4.7 (x)	8.2.79	Gigha	6.1 (x)	4.11.77
1978	Kieran	5.0	19.9.78	Risga	5.75	13.9.78
	Fingal	5.25	15.10.78	Danna	6.5	15.11.78
	Conon	6.0	16.10.78	Ulva	6.5	15.11.78
				Aida	6.75	24.11.78
				(Shona	6.5	died)
1979	Fillan	4.5	24.8.79	Eorsa	5.5	13.9.79
	Ossian	4.75	24.8.79	Gisla	5.5	19.9.79
	Conall	4.75	6.9.79	Tolsta	5.5	27.9.79
1980	Cowal	4.75	27.8.80	Morna	6.0	8.9.80
	Erin	4.5	27.8.80	Hynba	5.5	16.9.80
	Appin	4.25	9.9.80	Croyla	6.5	2.10.80
	Lorne	5.25	16.9.80			
	Bran	5.25	2.10.80			
1981	Mabon	4.5	18.9.81	Grania	6.0	18.9.81
	Nechtan	5.5	29.9.81	Forsa	6.75	16.10.81
	Brechin	5.0	19.10.81			
1982	Merlin	5.2	19.9.82	Petra	6.0	19.9.82
	Gregor	4.25	1.10.82	Hirta	6.0	29.9.82
	Eoin	4.0	3.10.82	Roma	6.5	29.9.82
				Tara	5.75	1.10.82
				Tulla	5.25	3.10.82
				Jura	–	7.10.82
				Nessa	–	7.10.82

No. of eagles imported: males 25; females 30
No. of eagles released: males 24; females 28
(x = mean value)

discovered Freya's cage but would not have realised that she was the female with whom she had shared her nest in Norway. She readily accepted food left at the cages but once Karla and Freya were set free (on 24th October and 1st November, respectively) neither cage held much attraction for her; I then established an additional food dump nearer the shore. By now all three eagles were finding deer grallochs (the discarded entrails of shot deer) which they readily located on the open hill. During November Karla and Freya were seen together but Loki must have left the island; her body was found on about 19th November beneath power cables in Morvern, Argyll, some 64 km to the south of Rum. Unable to recover the carcass we could never positively ascertain the cause of death.

The other two females remained on the island until the following February, and one was still present 2 months later. We received a reliable report of a Sea Eagle at Arisaig, on the nearby mainland, on 3rd August, 1976. One returned to Rum later that December, and was seen on neighbouring islands in April and May 1977. It was possibly a different bird – by that time in third-year plumage – sighted on Islay far to the south, on 24th February 1978, and 3 months later on the Antrim coast, just across the water.

In June 1976 we were licensed to receive 10 eaglets, which proved to be five of each sex (see Table 14). The first pair were released on 23rd November. Ailsa (whom I now believe to be a male) returned to the food dump within 2 days but another male, Fionn, was seen only rarely. Both were seen feeding at a deer carcass during the end of January 1977 and were still in the vicinity during March. A male, Cuillin was freed on 24th January 1977 and another, Brendan, on 28th February. Both were observed to use the food dump for 2 or 3 weeks. A female, Isla, released on 15th April was at once chased from the area by persistent mobbing of Hooded Crows and again, later that same day, by inadvertent human disturbance. The weather during the ensuing week was wet and misty, so that Isla was unable to find her way back. I suspect she may have starved, for exactly 1 month later her decomposing remains were washed ashore on the Isle of Sanday, 12 km to the west.

As soon as Iona had been released from her cage on 3rd May 1977 it had begun to rain. This prompted her to leap into a puddle and lie first on one side and then the other to splash herself with her wings. She soon became so waterlogged that she was unable to take off and had to seek temporary shelter to dry behind a large rock. For the next couple of weeks she frequented the shore and by 16th May had moved a few kilometres around the corner to Papadil at the southeast of Rum. Here, on 22nd May, I saw her indulge in yet another bath in the shallows of the freshwater loch. It is interesting that in recent years we have discovered eagle footprints in the sand along the shores of other lochs, suggesting that bathing may be a favourite occupation of certain individuals: it is just possible, however, that they may have been wading in the shallows looking for fish as Willgohs (1961) has described.

After the spring of 1977, sightings of birds released in 1976 became less frequent, although two were seen on the mainland nearby. On 7th August 1977 one was observed at sea between Skye and Rum: the yachtsmen saw it land momentarily on the water amid a flock of auks, possibly attracted by a shoal of fish.

The four eagles which were retained as captive breeding stock were tethered within sight of one another in a fenced enclosure (to keep them away from interference by cows, deer and people) on an open hillside near the cages. This site was to become a focal point of activity in subsequent years. In March 1977 Brendan began to steal food from the tethered birds so that a food dump was established there.

Because the tethered birds were never allowed to become tame they exerted considerable pressure on the materials restraining them. The first escape was achieved by Colla (Fig. 75), who broke the metal ring at the end of her leash. She flew competently from the outset and by dusk had made her way to the shore, choosing to roost on a low, accessible rock. I memorised the exact location and the easiest approach route before going to summon help. Armed with powerful torches and long-handled nets, Peter Corkhill and I set off silently in pitch darkness. We arrived at the appropriate rock where, at the last possible moment, we switched on the torch to find Colla dazzled in the beam. She took off just out of reach of the nets, but fortunately became confused in the beam and came to earth again only a few metres ahead. Immediately she was in Peter's net and within minutes was safely secured on a reinforced tethering wire.

Beccan (Fig. 76) escaped on 13th October 1976, circled competently several times and landed on a distant cliff face. She took off again and we soon lost her in the rough terrain. It was 3 days before she turned up again when, as with Colla, I was able to follow her to roost. Repeating the previous escapade, John Bacon and I approached within reach and dazzled Beccan in the beam. She took off and her weight broke the handle of the net. She was tangled in the mesh long enough for me to lay hold on her, so once more a fugitive was returned to captivity. On 6th December I noticed that Beccan had pecked part way through her leather jesses and I had to renew them.

Her ultimate moment of triumph came on 4th May 1977 when, at 1 year old, she again escaped, evading all attempts to recapture her. The dazzling technique proved ineffective because the summer nights were bright

Fig. 75. Third-year female Colla, one of our largest females, Rum, August 1978. (Photo: J.A. Love)

enough for her to detect any approach; this time she chose to roost in steep, rocky terrain which was difficult to cross quietly and in cover of darkness. Consequently I changed tack and laid out some bait to which were attached concealed nooses of nylon to entangle the eagle. Beccan was not to be easily deceived, however, and merely helped herself to food from one of her still-tethered companions! Somewhat embarrassed, I temporarily removed all other food so that this time Beccan had to feed at the bait. Once satisfied she tried to take off but found herself held. I sped to the scene and had almost reached her when she broke free. Two nooses had been pulled taut over her claws only, so that the nylon merely slipped off when she struggled. During June, Beccan wandered further afield and rarely returned to the tethered site thereafter. The tether dangling from her legs proved no hindrance and within a month Beccan was catching live prey regularly. Her escape was to prove a fortuitous event. The jesses and leash formed a conspicuous marker so that I received many reports to indicate her movements. She visited the neighbouring Small Isles and Ardnamurchan, on the nearby mainland, often returning to Rum to provide fascinating interactions with other Sea Eagles, as we shall see presently. Within 14 months all but a few centimetres of leather around one leg had dropped off.

Fig. 76. Head of juvenile female Beccan, Rum, July 1976. (Photo: J.A. Love)

In June 1977, with three eagles yet on tethers, a further four were imported from Norway (see Table 14). The first pair – Mungo and Vaila – were released on 18th October; they soared around together and suffered a brief attack from a passing Golden Eagle. They displayed a propensity to wander and only Vaila would return occasionally to utilise food dumps. Later that month Mick Marquiss arrived on Rum to demonstrate radio-telemetry techniques. We made up a small 20 g transmitter encased in epoxy resin. This radio was tied at the base of one of the female Gigha's central tail feathers. To prevent the 12-inch wire aerial whipping around and possibly snapping, it was threaded through part of the central shaft of the feather. Once the tail coverts were smoothed back into place the radio was invisible, and over the next few days we ascertained that it did not trouble the bird, nor had she attempted to remove it. Since Gigha herself weighed 6.1 kg, the additional weight of the radio was insignificant. On 4th November 1977 she was given her freedom. After several brief flights she landed beyond view, but I could pick up a signal using the portable receiving equipment; by taking a cross-bearing I could plot her approximate position. She did not move again before dusk and was at exactly the same spot the next morning. The heavy rain and mist inhibited my finding her with binoculars, although I could detect her brief sorties by the fading and surging of the signal; once she circled for about 15 minutes. The day remained wet and misty and she made only one or two short flights. During one which lasted 35 minutes she was attacked by two Golden Eagles but countered in the usual fashion by swerving to one side to present her talons. After another flight of 20 minutes, she disappeared over the crest of a distant hill. It was 2 days before I located her again, in a remote glen on the east of the island, but thereafter her movements became difficult to follow since she frequented high ground. On 16th November I had to terminate the trial but Gigha remained on the island for the rest of that winter.

In direct line of sight I had found that I could detect a signal from a distance of 5–6.5 km but on such a large and hilly island as Rum, such opportunities were rare. The best signals were received when I positioned myself on one of the high hills, but in so doing I sacrificed the mobility needed to obtain useful cross-bearings and to move position to follow the bird's signal. Often the smallest features could totally mask the signal. During the first 4 days that the eagle had been under continuous surveillance – 2 080 minutes in total – she made only 21 flights totalling 172 minutes; thus only 8.3% of her time had been spent in the air. The poor weather on 2 of the days may have grounded her more than normal and of course, being relatively inexperienced in flight, she may have been reluctant to spend much time in the air. Nonetheless, the observations do indicate similarities with those made by Brown (1980), who ascertained that African Fish Eagles spent 75–90% of their day perched. Brown also quoted a study of Wedge-tailed Eagles (*Aquila audax*) in Australia; their flights lasted from 6 to 90 minutes and in one of 35 minutes the bird covered an area of 4–5 square miles, indicating just how mobile eagles can be, compared with a man on foot.

A second brief trial was attempted in 1978 with two of the eight eaglets imported that year (see Table 14). A female, Risga, and a male, Kieran, were both fitted with tail-mounted radios and were released on 13th and 19th September, respectively. They were tracked in the vicinity of the food dump until the end of the month when both radios suddenly and simultaneously ceased to function. That month the rainfall was 66% more than average and we suspect that dampness had proved too much for the waterproofing of the radios. On 29th September, 16 days after her release, I found Risga on the enclosure beside the tethered birds, her plumage very wet and sodden. This hampered her clearing the surrounding fence so that I could grab hold of her. She was in good health otherwise but I retained her in captivity for several days, and after some good meals released her again on 2nd October.

Two males, Fingal and Conon, were set free on 15th and 16th October, 1978, and later three females – Ulva and Danna on 15th November and Aida on 24th November. At this time Beccan reappeared on Rum and came regularly to feed alongside the captive birds. It was often difficult to identify the various juveniles using the coloured rings but plumage feathers proved useful. Several birds had damaged or missing flight feathers or else showed distinctively pale or dark plumage. Being 2 years older than the others, Beccan was conspicuous; some of her body feathers were pale and worn but the newly moulted ones were fresh and dark, giving her a distinctive, mottled appearance. Her beak was quite yellow, as was that of the similarly aged male Ronan, compared with his captive sibling Sula; Ronan's voice also became more shrill in pitch at about this time. Over the ensuing year as Sula's beak became more yellow she retained a noticeable dark spot on each side; this was still detectable as a deep, brownish-yellow patch even in her fifth year, greatly facilitating her recognition in the field.

On 24th November, immediately following Aida's release, a new eagle appeared on the scene. She had a markedly pale plumage, almost white on her back and wing coverts. Several secondaries on each wing protruded an inch or two beyond the others and would seem to have been those retained from her first-year plumage. I eventually managed to distinguish her coloured rings which identified her as the female Gigha, released in 1977. Over the next month or two it was not unusual to see four or five eagles in the air together, or coming in to feed alongside the tethered birds. Several roosted on a large rocky knoll near the food dump. Beccan was a central figure in this aggregation; she often flew wing tip to wing tip with Gigha, several times swooping to present talons, once turning a complete somersault, like a Raven but with her wings still fully extended. On 4th December 1978 I watched Beccan attempt to take food from the tethered female Sula, who defended herself vigorously. Instead, Beccan approached the male Cathal, who meekly permitted her to feed. Immediately Beccan took off she was pursued by the juvenile male Fingal, who screamed hungrily. Both landed on a nearby slope but Beccan took off again at once, leaving the meat for Fingal; she procured for herself another portion from the food dump. The following day Fingal was again seen to pursue Beccan, screaming and swooping up from below in an attempt to snatch the prey from her talons. Eventually he succeeded in this but as soon as he landed to feed, he was dispossessed by three other youngsters, one of which consumed the food. Such interactions became more frequent but the eagles were showing less attachment to this site and wandered to other parts of the island. In early January 1979, during a period of exceptional frost and snow, Beccan and at least three other eagles turned up on one of the neighbouring islands where they were seen to feed upon a sheep carcass. Gigha and Ulva remained on Rum to be rejoined 2 weeks later by Beccan and the obsequious Fingal. I once watched Sula leap at Beccan angrily to keep her at bay, but after two or three such attacks Sula had moved sufficiently far for Beccan to pounce on her food unmolested. Beccan gained height with her ill-gotten prey, inducing in Ulva some unrewarded food-begging behaviour. On 6th February, both Cathal and Sula successfully fended off Gigha, but a wily Hooded Crow managed to make off with a small fish. Immediately Gigha gave chase and the 'Hoodie' dropped its prize for fear of its life. Gigha retrieved the fish and rose to a height of 100 m to hover into the wind and consume the fish directly from her talons (Fig. 77). In this way she evaded a half-hearted attempt by the juvenile Kieran to snatch it from her.

The next day Fingal was also seen to retrieve a scrap dropped by a Hooded Crow, but Conon had to challenge both Sula and Cathal before he won some fish, which he hastily snatched in his beak; he took to the air and deftly transferred his prize to his talons. On 19th February, while I was replenishing food supplies at the tethered birds, Conon appeared above holding a clump of grass in his talons. He flew around clutching this for some time, perhaps an act of impatience and frustration as my presence denied him access to the food dump. I retired to watch him swoop down on Ronan, hardly pausing as he snatched a talonful of small fish. He had also torn up a quantity of dried grass, so hovered into the wind to pick out and swallow some of the fish; he carried out the operation with less accomplishment than had earlier been shown by Gigha – at a lesser height and dropping several scraps in the process. When he swooped low for more he induced much excited calling from Sula on her perch.

One day in February I was watching four eagles soaring together above the roost when all of a sudden two Phantom jet aircraft zoomed up the glen at about 600 m. None of the eagles reacted to the noise, nor was in much danger, but the possibility of a bird strike – the first ever encounter between NATO and British Sea Eagles – was a sobering and disquieting thought.

It was encouraging to see the continued presence of three or four of the seven eagles released in 1978 after 5 months of freedom – a period which would doubtless be the most hazardous for young and inexperienced birds.

Fig. 77. Eating prey in flight.

They undoubtedly benefited from the example set by the older birds such as Beccan and Gigha, who lent a certain cohesion to the group. The continued attraction to the food dumps was further encouraged by the three eagles tethered nearby. By this time, however, we were experiencing pressures from several quarters to release them. On 8th February 1979, the $1\frac{1}{2}$-year-old male Cathal was liberated and although the next day he visited the food dump we never saw him again. I suspect that, like Isla in May 1977, he had starved within a month or two since he failed to utilise the available food supplies. In mid-August his skeleton was discovered on the beach at Sanday – strangely enough at the exact spot where Isla had been found a year earlier – a bizarre quirk of currents and circumstance. Loki had similarly succumbed 2 months after her release (although apparently by accident) and these three ringing recoveries confirmed the hazards which the young have to endure early in their freedom.

The second long-term captive, Sula (Fig. 78), was given her liberty on 6th March 1979 and Beccan observed the event from above. As soon as Sula took to the air Beccan swooped on her, forcing her to flip over and show her talons – quite an accomplishment on her maiden flight after nearly 3 years' confinement. After several minutes in the air she landed on a grassy slope with Beccan by her side. The latter took off, circled and returned two or three times but Sula refused to be moved, twice jumping at her aggressively. The 1-year-old female, Ulva, then appeared on the scene, so Beccan pursued her instead; momentarily they touched talons in mid-air before alighting beside Sula. When Sula did finally take to the wing, Beccan followed her and coerced her into touching talons, perhaps in threat. Sula landed beside a

Fig. 78. Third-year female Sula, with characteristic speckled plumage of immature, and beak turning quite pale, Rum, January 1979. (Photo: J.A. Love).

small pool where she proceeded to bathe. Beccan sat patiently nearby, but was forced to take wing by a persistent Ulva; however, she was reluctant to show much interest in the young female. Eventually Sula emerged to dry off on a knoll nearby. I found her at the same spot the next day but by the third day she was testing her flight capabilities in the strong onshore breeze. Several Hooded Crows mobbed her and one even succeeded in momentarily tugging at her extended wings.

It was at this time that the individual Shona was placed on tethers where she quickly benefited from the example of the still-captive male Ronan, even attempting to flutter weakly on to the roof of her shelter. On 24th April 1979, once we were sure that she had settled on her tether, we released Ronan (Fig. 79). Once or twice during the ensuing weeks he returned to feed alongside Shona but obviously rejoiced in his new-found freedom. He

Fig. 79. Third-year male Ronan, Rum, January 1979. (Photo: J.A. Love)

wandered extensively and by early June was seen regularly near the beach at Kilmory, at the north end of Rum. We found several gull carcasses in the area and Ronan seems to have been responsible too for killing two Eider ducks as they innocently incubated on their nest. Ronan had developed his predatory skills at an early stage, even before his release, for one day he had snatched, killed and consumed a Hooded Crow which had foolishly strayed within his reach. His notoriety became firmly established, however, when on 23rd June he killed a Red Deer calf. This had been born only 2 days earlier and was being observed by one of a team of scientists studying the Red Deer at Kilmory. After a heavy squall an eagle emerged from the hollow where the calf had been lying. Upon investigation the scientist found the calf dead, its ribs opened up and one foreleg torn out and eaten. The carcass remains weighed 5.2 kg – some 1.8 kg less than when it had been weighed and marked only a few hours earlier. While in captivity Ronan had weighed about 5 kg, so at this single meal he had consumed nearly 40% of his own body weight. The calf had been healthy and of average weight, but its mother was known to be particularly inattentive and a poor parent. Skinning the carcass revealed no talon marks nor were there any signs that it had struggled; the wound, where the eagle had grasped its prey and where it had fed, must have been fatal.

On 20th June 1979 I flew to Norway with the RAF to receive from Harald a further six eaglets (see Table 14). They proved to be three of each sex, though their ages varied from 6 to 8 weeks. All fledged successfully and the first two males, Fillan and Ossian, were freed on 24th August 1979. Within 4 days they were returning to feed at a deer carcass which had been left nearby. Sula later teamed up with them and all three were seen feeding at a deer gralloch on the open hill some distance away from the food dump. The remaining four eagles were liberated during September; the male Conall was released on the 6th, and the females Eorsa and Gisla on the 13th and 19th. They all established a routine of coming to the food laid close to where the last female, Tolsta, was being kept; she was released on 27th September. Both Ronan and Sula 'adopted' the juveniles, assuming the role that Beccan had played in previous years, her visits now being infrequent. On one memorable day I watched seven eagles in the air together. Several attempted to talon-grapple, but only the pair of older birds could successfully interlock, cartwheeling out of the sky amid excited screams. Both Ronan and Eorsa were seen to pursue Ravens from the food dump to pirate scraps from them.

On 18th September 1979 the invalid Shona died, so that there were no longer birds tethered to act as decoys. As a result, the 1979 releases forsook the food dump at an earlier time than had those of previous years, but they did remain in the company of the older pair. They were seen to locate grallochs and to feed at a deer carcass and on a dead seal washed ashore. I shot three goats for them but these were ignored, suggesting that there might have already been an abundance of carrion available. During the summer Sula acquired an oily patch on her breast, presumably from having killed some Fulmar chicks. Her diet of carrion thereafter, together with her constant preening, soon rid her of the oil. She now appeared firmly attached to this corner of the island, while Ronan was once more seen to frequent Kilmory, where he accepted some fish I had laid out near the beach. He was actively hunting gulls and on one occasion I watched him employ both a heavy squall and the sun at his back to launch a surprise attack, albeit an unsuccessful one. He also fed at the carcass of a stag near the shore, but simultaneously evaded my attempts to photograph him from a hide which I had placed nearby. At times he would move along the coast, even being seen over the village of Kinloch.

During the summer of 1979, Beccan forsook the food dumps to take up a territory on sea cliffs nearby, an instinct developed at this same age by Sula and, to a lesser extent, by Ronan who, at a later date, was to pair up with

Beccan. On 10th January, having apparently spent the summer on an island 35 km to the south, the very pale-plumaged female Gigha reappeared on Rum. In mid-March she still accompanied Sula. In early April Sula entered her annual moult. Twice during that spring her footprints were found on the sandy shores of two freshwater lochs, where she had presumably bathed. One or two reports of the 1979 releases were received during the summer. Perhaps the most extraordinary was that of the eagle seen by my colleague Nigel Buxton near Tolsta in Lewis on 26th June. Being able to approach the bird closely he could distinguish its two red rings on its left leg. Imagine my astonishment when I consulted my notes to find that the eagle Nigel had seen was the very one that I had named Tolsta!

On 20th June 1980, Martin Ball and I had flown to Norway to receive from Harald Misund eight eaglets (see Table 14). Their collection from eyries had been filmed by a BBC television crew. Hugh Miles remained to shoot sequences of adult Sea Eagles at the nest, while another crew accompanied us to Rum to cover the later stages of the operation – tethering, feeding, etc. I released two males, Cowal and Erin, on 27th August so that when Hugh visited Rum a few weeks later he was able to film them from a hide at the food dump. He also shot the release of a female, Morna, on 8th September, and that of the male Appin the next day. Not to be left out, Sula too obligingly performed for the camera that day. The completed film, with some fine sequences of Sea Eagles catching fish in Norway and a commentary by David Attenborough, was screened a year later.

The female Hynba and the male Lorne were set free on 16th September, and I retained until 2nd October a female Croyla and a male Bran. Hynba was the most ready to return to the food dump and on 10th October I found her perched confidently on the shelter to which a month before she had been tethered. Two weeks later I watched her approach the food dump which that day included a Greater Black-backed Gull which I had shot earlier that day. Although interested in the gull she was reluctant to come near it. After an hour had passed she approached cautiously. After several minutes she leapt tentatively forward but only brushed it with a talon, as though to ascertain that it really was dead. The next time she swept it up, and several metres distant finally plucked and ate it. On several other occasions I had been aware that some eagles had displayed similar nervousness at approaching a dead bird, preferring if possible to accept instead meat or fish.

On 27th October, Hynba had a bath in a shallow freshwater pool near the shore at Harris. She flew to a nearby fence post to dry off and preen, but first balanced precariously for a moment on the wire. Early in November Gigha again reappeared, having been absent as in previous summers. On 24th

November I observed eight different eagles at Harris and, as is to be expected, the interactions during the ensuing weeks were many and various. On one amusing occasion I was watching several of the juveniles near the shore. One was flying lazily along the raised beach where it was stooped at by a bold Hooded Crow. Absent-mindedly the eagle flipped over to threaten with bared talons but sacrificed so much height that, being only 2 metres up, she crashed to the ground in an untidy and ignoble heap. Once regaining her composure, the embarrassed eagle vented her annoyance on one of her peers perched innocently nearby.

In June 1981 we learnt that the RAF had to bring forward our flight which, together with its being a poor and disrupted breeding season in northern Norway, resulted in Harald's being able to collect only five eaglets (see Table 14). Three were almost fledged while the other two were barely half-grown, but all were reared successfully and the two youngest – a female, Grania, and a male Mabon – were the first to be liberated, on 18th September 1981. Another male, Nechtan, followed 11 days later and a female, Forsa, and male, Brechin, were released in mid-October. Sula had never forsaken the vicinity all year and, as if on cue, appeared as soon as Mabon had taken wing. She remained with the juveniles near the food dump, with three other eagles released in previous years also turning up. The first was nearly adult, possibly the female Gigha, another was established by its red ring as a male set free in 1978 and the third I was able to identify positively as the familiar female Hynba from 1980. It was especially gratifying to know that the nine eagles chose to remain on Rum that winter. We received reports of others elsewhere but sadly, in late April 1981, we had learnt that the male Erin (released on 27th August 1980 and soon after filmed by the BBC) had been found dead in Caithness. Analysis of his carcass revealed that he had been poisoned by having fed on a hare carcass laced with the lethal poison Phosdrin. Otherwise the bird had been in good condition and was of a healthy weight. In November 1981 we were notified of our fifth recovery, from the northwest of Skye – perhaps the greatest loss of them all. The bird was a female, Vaila, which had been released in 1977: we had received no news of her since and she was by then almost old enough to breed. She had been dead several months so that little remained of her carcass to permit analysis, but a dead gull lay nearby and both birds lay a short distance from the carcass of a sheep which had been pulled to a conspicuous position on top of a knoll: we suspect that Vaila had been poisoned. At least two other Sea Eagles were seen in the area so, as a precaution, our contact buried the sheep.

Our sixth recovery came in May 1982. The skeleton of the 1980 female Hynba was washed out from a ditch on Canna during a period of prolonged

Fig. 80. First-year Sea Eagle, newly released on Rum and showing its numbered, patagial wing tag. The bird (named Gregor while in captivity) is scavenging along the shore, Rum, October 1982. (Photo: J.A. Love)

and heavy rain. She had been released on 16th September 1980 and remained dependent upon the food dump throughout that and the following winter. In December 1981, during severe and heavy snow conditions, she disappeared. Hunting may have become difficult for her, especially as I was unable to replenish the food dump because the access road was blocked, and she seems to have moved on to Canna. It is possible that in a weakened state she leapt into the ditch to feed upon a dead sheep and was unable to get out (a technique used to advantage in the Highlands to trap and kill eagles last century); or perhaps she just starved to death in the cold weather.

On 22nd June 1982 10 eaglets were flown to Kinloss by the Norwegian Air Force. All fledged successfully and while in captivity served to attract back to the vicinity the 1976 female Sula in the company of a 1979 male and the 1980 female Croyla. The female Petra and the male Merlin were released in mid-September and utilised the food dump for several months. The others were set free during the next 3 weeks – six more females and only two males.

These eight were fitted with patagial wing tags, each bearing a conspicuous digit. Although rather obtrusive these tags (orange on the right and red on the left wing) greatly facilitated individual recognition. A male, Gregor, bearing the number 6 (Fig. 80) immediately forsook the vicinity and turned up 6 days later on the shore at Kinloch some 4 miles to the northeast. He was later seen at the north of the island but soon afterwards was swept far to the north during a period of easterly gales. On 19th October Nigel Buxton contacted me to report Gregor's presence north of Stornoway in Lewis.

Thus, up to the end of 1982 – the eighth season of our reintroduction project – a total of 55 eaglets have been imported (see Table 14). Of these, 52 were able to be released into the wild on Rum – 28 females and 24 males. Four females and two males have been recovered dead subsequently so that our sex ratio is close to 1:1. It is hoped to import further young over the next 2 years to establish a strong pioneer population from which we hope to achieve the first breeding successes.

10 Recolonisation

Sometimes eagles' wings,
Unseen before by Gods or wondering men,
Darken'd the place.
John Keats (1795–1821)

The 'hacking' procedure employed in the reintroduction project has served to implant Sea Eagles into the vacant habitat of the Hebrides. Once liberated, the youngsters usually return to the vicinity of the cages/tethers to utilise the food which is provided for them there. Despite the fact that the species is more confiding towards man than is the Golden Eagle so that – as in Norway and elsewhere – it is often reasonably approachable, the coloured rings have not proved easy to read in the field. Thus, unless a bird possesses some conspicuous plumage characteristic or a broken feather, we cannot always identify individuals. (Tags used in 1982 are proving better.) We do know, however, that only seven birds (four females and three males) have failed to return. Two of them – the 1976 female Isla and the 1977 female Vaila) have since been recovered dead. Vaila's loss was all the more sad as she had proved her ability to survive in the wild for 4 years. Clearly we cannot dismiss all 'non-returns' as having died. It is significant too that a Bald Eagle which was 'hacked' into New York State in 1976 was not seen for 4 years but then suddenly returned to breed.

In all, six Sea Eagles (four females and two males) have been found dead. The only factor common to all six is that they were all below average weight for their sex – from 2 to 10% below. This may not predispose to an early demise, however, since Sula too – still alive and well at the age of 6 years – was 6% lighter than average. Doubtless other deaths have gone unreported. 1980 Erin and 1977 Vaila were poisoned and if 1976 Loki did not kill herself against power cables she most probably died at the hands of man. Vaila, probably Erin (who at the time of his death was still a healthy weight) and possibly Loki too, may otherwise have been alive still. 1976 Isla had failed to feed at the dump while 1977 Cathal had returned to it only once. Both had been released in the spring along with six other eagles – a known mortality of 25%, compared with only 12% amongst the autumn releases. If this difference is real and not due merely to the small sample size, it is difficult to account for it. The main mortality amongst the island's Red Deer occurs in late winter/early spring, providing a sudden abundance of carrion which, together with the increasing daylength, may encourage the young eagles to wander prematurely from the security and reliability of the food dump. The 1980 female Hynba seems, on the other hand, to have remained dependent upon the artificial food so that in her second winter when forced to disperse during severe weather, she was unable to provide adequately for herself.

Isla, Cathal and 1975 Loki all died within about 4 months of their release, so this is obviously a critical time. Thus it may be useful to consider two other categories of birds – those which have been seen up to 4 months after their release and those which are known to have survived in excess of this time period. The former group includes nine males and six females (excluding the 1982 imports who at the time of writing have not been free long enough to qualify). Both Loki and Cathal are among them, and it is likely that others may also have succumbed. Observations would indicate, however, that this is an adequate time interval for the eagles to acquire hunting skills.

In our final category – long-term survivors – we can confidently include no less than 21 eagles, exactly 50% of all those released (again excluding the 1982 imports). Three – Vaila, Hynba and Erin – are known to have since died, but we should consider that the survival prospects of the remainder are indeed high.

It is difficult with such small samples to identify any physical attributes (such as release weight) or external factors (such as climate) which may influence survival. The Sea Eagle is a gregarious species so that sociality must be important. During the first 3 years of the project, only 13 birds were released, at widely separated intervals. Including the four retained in long-term captivity, 29 were released in the next 4 years. By then the desirability of holding back 'decoy' birds near the food dump was recognised so that the new releases were encouraged to remain in the vicinity and to team up together. Also several of the older birds were returning to the island and not only tolerating the juveniles' company, but also acting unwittingly as foster-parents: one or two were even seen to yield up food to them. Thus, in these later years we have fewer deaths and 'non-returns' and can recognise more long-term survivors (despite there being less time for the latter birds to be detected and thus to qualify).

As the number of eagles being released on Rum has been sustained over the years (Fig. 81), we have received more and more reports, not all the birds of course being on Rum. These have permitted some conclusions to be drawn regarding their dispersal. A total of 739 observations derive from Rum (up to the end of 1981) and a further 230 from elsewhere. All but 18 have been sighted within an 80 km radius of the release point – 98% of all sightings. This demonstrates how our eagles are disinclined to wander far (Fig. 82). Our most distant report is from Fair Isle, 350 km to the northeast of Rum, although in 1982 Sea Eagles have been reported from the south of England; if indeed they are 'our' birds and not true migrants from the continent, they may have been induced to wander south during the harsh 1981/82 winter.

In examining the timing of these reports we note, not surprisingly, that on Rum most tend to be in the months following releases (September to December). Not all of these refer to newly-released eagles, however, as individuals from previous years seem to be attracted back to the island by the presence of the new imports in captivity. Nearly 75% of sightings on Rum of birds 1 year or older are also during the months September to December or January. By the spring and early summer the eagles begin to disperse. Reports from Rum (as with the Fair Isle releases in 1969) become less frequent in the summer: my observations indicate that this pertains also for Golden Eagles. Perhaps while the birds are in heavy moult they prefer to remain more secretive. Any of the sightings we do receive away from Rum, however, tend to be in the summer months, presumably when locals, tourists and bird-watchers alike are most active.

In the spring of 1981 we accumulated enough observations within a short space of time and from such scattered localities for us to deduce that some 20 to 25 of our eagles were still accountable – out of a total of 37 then freed. By the following summer we had released 42 and still knew the whereabouts of over 20. Few of these individuals could be identified positively so it is likely that there may be more than we tentatively suggested, and that others had escaped detection. Thus we are inclined to the view that survival is in excess of 50%. Only 15% (six birds) are known to have died, a mortality much lower than may prevail elsewhere in Europe (Cramp & Simmons, 1980).

These encouraging observations show that not only has the survival of the young been high but that the environment into which they have been

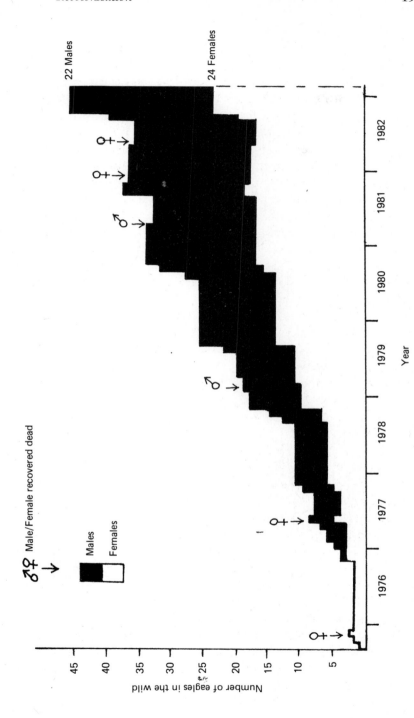

Fig. 81. Number of White-tailed Sea Eagles released from Rum (1975–82) and which are potential survivors.

Fig. 82. Sightings of White-tailed Sea Eagles released on the Isle of Rum 1975–81. Large circle has a radius of 80 km from Rum (star) and contains over 95% of all sightings; other confirmed sightings are marked by a dot, possible reports by a question mark. The asterisk marks the only recovery of a dead bird to have been found outside the circle. The data for 1982 thus far received indicate that, following the severe winter, some young Sea Eagles may have wandered further afield than hitherto.

released is still eminently suitable for them. Some fish species may not be as abundant as formerly but we do have evidence that the eagles are indeed catching fish. Two birds have been observed hunting at sea presumably for fish, whilst two others have been seen carrying fish in their talons – one bird just off the shores of Rum and another on a freshwater loch well inland. Fish bones have been found on several cliff roosts known to be frequented by the Sea Eagles, while in March 1981 Roy Dennis watched the 1980 female Morna show an active interest in an otter and her two cubs fishing in Loch Torridon.

We have much more evidence of seabirds being killed – auks, Shags and gulls regularly. Bob Swann watched one eagle landing momentarily on Kittiwake nest ledges to lift chicks. On Rum, several Manx Shearwaters have fallen prey and the eagles are regularly seen chasing Hooded Crows and Ravens. One mobbing Herring Gull was seen to be snatched in mid-air. Fulmars are taken, perhaps mostly in flight, although in the autumn of 1980 Sula was found to have oily patches on her breast feathers. Fortunately this was slight compared with the unfortunate Fair Isle release who had been catching young Fulmars in the nest, and Sula rapidly cleaned it off. On some sea lochs the eagles regularly hunt wintering ducks. Ronan killed two innocent Eider as they incubated; he also killed a two-day-old Red Deer calf on Rum. On an island in Argyll two immature Sea Eagles were seen attacking a hind and her calf, but by rearing up on her hind legs the deer managed to defend her young one. Mountain Hares are common prey on that island; elsewhere, rabbits, especially those weakened by myxomatosis, are favoured.

Carrion is of course an important source of food for the eagles: Red Deer and Feral Goat carcasses occur on Rum, and Sula was once seen to feed upon a putrid Grey Seal lying dead on the beach. Dead rabbits are easy pickings and once a Grey Lag Goose carcass provided a meal. Erin was poisoned when feeding on a hare carcass and Vaila at a dead sheep. It would not be surprising if the eagles lifted an occasional dead lamb, and dead sheep are a lucrative source of carrion for them in the Highlands and Islands. On one sheep walk now regularly frequented over six seasons by Sea Eagles, no lambs have disappeared unaccountably: the shepherds tend their flocks well and indeed take an active interest in the progress of 'their' eagles.

It is interesting that to date, despite regular media coverage, the project has attracted little adverse criticism. The two poisonings which have taken place were probably directed (still illegally!) at foxes or Golden Eagles, but the fact that this took place at all and that more poison baits are being detected of late, are causes for very great concern. The Sea Eagle, as we have seen, is, by its love of carrion, especially vulnerable to this dastardly habit, but for this very same reason is the least culpable.

Another more insidious poisoning threat might also be predisposed towards Sea Eagles – toxic chemicals. In recent decades coastal pairs of Golden Eagles have experienced difficulty in breeding, with success much reduced when compared with inland pairs. On Rum, for instance, an average of only 0.27 young were produced from each breeding attempt (Corkhill, 1980) compared with 0.7 on the mainland (Everett, 1971). While the coast may never have been an optimal habitat for Golden Eagles, Corkhill implicated DDT and PCBs acquired from seabirds upon which the Rum pairs sometimes feed, although this may be only because of the lack of Rabbits and hares on the island. We may have cause for anxiety that the Sea Eagles also take seabirds, but might derive comfort that they may be overestimated in the birds' diet. Using more accurate methods of prey determination, Wille (1979) revealed that up to 90% of the diet in Greenland may in fact be fish. Furthermore, in the winter months and during their years of immaturity the Rum Sea Eagles feed to a large extent on carcasses of herbivorous mammals from which (since the ban on dieldrin sheep-dips) contamination risks are minimal. Although the waters around Rum, emanating from the Irish Sea, may – by British standards – be fairly polluted, they do not compare with the Baltic, and there are indications, albeit yet indirect and circumstantial, that the pollution levels around the island may be declining. Another coastal predator, the Peregrine, which had ceased to breed in the late 1960s, made a comeback to Rum in 1980 to rear one chick: now two pairs may be in residence. Similarly, in the 6 years following Corkhill's assessment of the Rum Golden Eagles' breeding success, their reproductive rate has doubled from 0.27 to 0.43 young per breeding attempt per year (Love, unpublished).

We might consider side-stepping any pollution doubts by releasing Sea Eagles elsewhere instead. In 1979 I reviewed the historical evidence, together with the suitability and practicality of other potential release sites. Among them, Shetland emerged as a conspicuous favourite. Its prime disadvantage was its remoteness, which would restrict the opportunities for the eagles to disperse and establish elsewhere in Scotland. Since they are gregarious birds we and our Norwegian colleagues inclined to the view that it would be unwise to dilute our efforts to date. The presence of eagles already released enhance the survival prospects of the new releases. Their flocking at favoured hunting grounds discourages dispersal so that pair formation might ultimately prove more successful. At best we should consider a release point which could overlap with the range of the birds already liberated on Rum, but this would still leave them within the sphere of pollution in west-coast waters. Until such time as resources and a surplus of imported eagles present themselves, any releases at an alternative venue have been left in abeyance.

Within 80 km of Rum, several localities have proved particularly attractive (see Figs. 83 and 84) – often holding groups of six or more birds, but in the interest of the birds' welfare I am going to be deliberately vague about the exact locations.

At the first site, some miles from Rum, we first received reports in 1978: a juvenile in September and two other possibles at around the same time. 1979 yielded a further 15 reports, again all immatures, and one was recognised as having been released the previous year. In three cases, two eagles were seen together and once they engaged in aerial display. We received 33 sightings in 1980 with as many as four immatures being seen together, and for the first time in the area an 'adult' with a conspicuous white tail was spotted.

Clearly a more concerted effort was desirable, so in 1981 and again in 1982 observers were employed by the RSPB to search the area thoroughly from February to August. In all they were able to recognise 15 individuals in 1981 and seven the following year. These birds hunted along the shore but were also seen to be especially fond of two inland areas. A small conifer plantation began to be used as a roost by several immatures in 1981. The following May it was used by a 2-year-old, rather pale-backed male and a 3-year-old female, until a few weeks later they were scared off by casual timber operations. After discreet approaches to the foresters, further felling was suspended. The two immatures remained together as a pair and, on occasions, were seen at the other inland site with a juvenile and four mature Sea Eagles. Three of the latter were fully adult and one had a black ring indicating it to have been an eagle released in 1975, the first year of the project. Two of the adults were seen copulating on 29th May 1981 on a rock ledge and the following spring indulged in further aerial display and talon-grappling. At times a 4-year-old (subadult) female would interfere and once she attempted to talon-grapple with the young pale-backed male. In April and May, the adults carried some sticks back to a cliff a few miles from the sea. The ledge was nearly 5 m long and 1 m wide and by mid-April/May had about a dozen sticks on it, arranged in a rough rectangle about 2 m × 1 m. This crude nest, under an overhang, was littered with droppings, down and food remains. At the end of May other sticks had been added with more lying on the slope beneath, but in a spell of wet weather the ledge became waterlogged and was abandoned. It had been rather accessible, so we were pleased to see that the pair favoured a better ledge some distance above the old one. It is hoped that they may build a new nest here or in the vicinity and they certainly seem poised to make a more determined breeding effort in 1983.

The third adult in the area may have been the one who the previous year had carried some sticks to a cliff ledge on the nearby coast. She may yet find

Fig. 83. The coast at Harris with the Cuillin of Rum behind – a haunt of the Sea Eagles upon release. (The mausoleum in the foreground is the burial place of the Bulloughs, former owners of the island.) Rum, January 1980. (Photo: J.A. Love).

a suitable mate among the immatures often seen in her company. A fourth locality was used regularly in 1981 by two adults but no trace of them could be found the next season. It is possible that they attempted to breed but at a locality nearby unknown to us. Certainly this and the coastal site used by the lone adult are remote and little visited by humans.

We know of two other lone adults present in the west of Scotland in 1982, one long-established at an inland site where it regularly displays to the local Golden Eagles. We do not know the sex of the bird but it has lately built a crude eyrie of sticks in a low cliff.

On Rum itself, 6-year-old Sula has become established but, although she is frequently in the company of various immatures, she has yet to attract a mate of breeding age. In 1980 I built a rough nest of sticks on one of her roost ledges in the hope that it might stimulate her to build it up for herself. It is still used as a roost by Sea Eagles but now Sula herself seems more attracted to a recently-vacated Golden Eagle site a mile or two distant. These Golden Eagles were never successful in their breeding attempts and latterly it may have been the persistent presence of Sula and her consorts which finally induced them to move away altogether. As a species the Golden

Recolonisation

Fig. 84. Seabird colonies on the east coast of Rum; the haunt of feral goats and a few Red Deer, and often frequented by the Sea Eagles, Rum, May 1980. (Photo: J.A. Love)

Eagle is the more aggressive and in the early years of the project I frequently witnessed their attacks on the Sea Eagles. Once a dive was so violent that I momentarily feared for the victim's life! In the last six years of the project, however, I have seen only seven such attacks compared with 15 in the first 18 months – a reduction from 0.8 to 0.1 encounters per month. It seems now that the Golden Eagle is more tolerant of its new neighbour.

It might also be that the two species are less liable to cross paths. It is only in recent decades, as the Golden Eagle has increased or else expanded its range, that it has tenanted coastal sites – some of them having formerly been occupied by Sea Eagles. In countries such as Norway, where both species are still to be found, the Golden Eagle frequents inland hill areas, feeding upon grouse, hares, etc. The Sea Eagle, on the other hand, is a bird of the coast or freshwater lakes, where it subsists on fish, waterbirds and carrion. That this situation may now be in the process of being restored in the Hebrides is indicated by three coastal pairs of Golden Eagles on Rum. The first I have already mentioned. Two other pairs have moved to new eyrie sites, one of them well inland. Both old sites were lately frequented by our young Sea Eagles and in September 1979 two of them – at an age of $3\frac{1}{2}$

During that winter I witnessed much excited calling, aerial display and talon-grappling. On at least one occasion one bird was seen carrying sticks in its talons. This courtship behaviour had intensified by the following spring and several sticks had been deposited on suitable ledges. I could trace no further developments that summer and when the cliff vegetation blossomed these ledges became rather overgrown. One or two were retained as roosts and on 20th August 1980, for instance, we disturbed one of the adults from the very ledge where earlier that year I had found several sticks. Although its beak was fully yellow and its plumage the characteristic grey-brown of an adult, some of the tail feathers retained dark tips: it was a male and in heavy moult. Suddenly to our left a second eagle appeared, very much larger and obviously a female: it seemed to be the same age but in a more advanced state of moult, its tail almost regrown fully and quite white. Both soared around the cliffs side by side before being lost from our view. It was only later that I managed to approach to read the coloured rings of the male; this revealed him to be Ronan – imported in 1976 and released in 1979. By the short length of leather jess on one leg the female revealed herself to be Beccan. (At the time of the escape from captivity in May 1977 I had thought her to be a male.) Several ledges were being used as plucking posts and I found the remains of young Fulmars, auks and gulls. One of the eagles flew some distance out to sea where a flock of Guillemots were fishing.

During the winter of 1980/81, aerial courtship continued and intensified. The pair were strongly attached to the highest section of cliff and where Golden Eagles had had an old eyrie. The ledges were inaccessible and difficult to view from above. We could only speculate from their activities that a nest had been built and eggs might even have been laid. In early June, from our boat 'Rhouma' just offshore, we could make out a structure of sticks under an overhang about 150 m above the sea, and about a quarter of the way down from the cliff top. The eagles could not have chosen a more secure nest-site. Perhaps it was here that the Sea Eagles had nested a century before.

I soon found a spot at the foot of the cliffs where I could view the edge of the nest from a safe distance but by then it seemed to be unoccupied. Three prospecting Fulmars were resting on the edge, something they would not have dared do had the eagles been in residence. During the remainder of the summer the eagles became more secretive and only rarely could they be flushed but from a section of the cliff further to the west; they were in heavy moult. Sometimes they interacted with the three other immature/sub-adult Sea Eagles in an adjacent territory.

Our expectations were high for 1982 but for some reason Beccan and

Fig. 85. Adult White-tailed Sea Eagle.

Ronan were rarely to be seen. Despite a thorough examination of the cliffs from above and below, and from a boat, I failed to locate a nest. Last year's site showed no signs of occupation. Later in the summer the pair reappeared to demonstrate their continued presence on the territory.

Any sense of disappointment in 1982 is, however, at once eclipsed by the knowledge that we now have two (and possibly three) mature pairs in occupation of territory. Three other adults have established themselves but have yet to find mates of breeding age. Indeed we have encouraging reports that of the 16 eagles released from the first 3 years' importations, no less than 10 have survived to adulthood.

We are confident that the young eagles are being released into a secure and suitable habitat so that their survival prospects are high. Barring unforseen disasters, the reintroduction project seems well on the way to success. It is a tremendous thrill after a lapse of 70 years to have Sea Eagles, their white crescent tails gleaming in the sun (Fig. 85), soaring amongst the cliffs that were once their home – and that have remained their birthright.

POSTSCRIPT

1983 was an exciting year! A Sea Eagle nest was found on a cliff on 6 April with a pale-headed female incubating. By 9 April, however, a second female – distinguished by her dark head – had appeared on the scene, and was observed to copulate with the male on a ledge 200 m from the nest. The male and both females shared incubation but gradually the second female took over completely from the first. On 14 April the nest was found to contain two intact eggs, together with the fragments of at least one other egg, presumably from another clutch. Incubation continued normally until the nest was deserted on 17 May: 2 days later it was found to contain one damaged egg (empty of its contents) and fragments of another egg on the rim of the nest cup. The clutch may have been damaged on 9 April when the first female had attempted to block her rival as she walked along the ledge to the nest: the second female edged past her however and with careless enthusiasm jumped into the nest cup to commence incubating.

On 29 April a second pair were found with a nest. The male displayed dark tips to his otherwise white tail feathers (although this is not always a sign of immaturity). The female however showed much less white on her tail and with her speckled back appeared altogether more immature. The nest contained one egg on 2 May and the female was seen incubating; by 15 May however the nest had been abandoned and the egg was broken.

Thus, both pairs failed in the breeding attempt in mid-May following a spell of bad weather, and in neither case could the circumstances be considered normal. One female was usurped by another during the incubation period while at least one of the second pair was immature. It is perhaps unfortunate that these first breeding attempts should have been made in a season which also proved to be a poor one for Golden Eagles in the Highlands (R.H. Dennis, J. Watson and D. Langslow, personal communication).

Nonetheless it is encouraging that the Sea Eagles have succeeded in laying eggs (three clutches in all!) and that at least one clutch was incubated for the full term. Full-time observation and wardening at the nests was made possible by generous financial aid from Eagle Star Insurance. In 1983 too we found a hitherto unknown eyrie which appeared to have been built and perhaps used the previous year: we know of a further three, perhaps four, pairs of adults which have established breeding territories. We are optimistic therefore that 1984 will produce the first Scottish-bred young. On 16 June 1983, 10 more eaglets (collected by Harald Misund) were flown to Scotland by the RAF, for release on Rum in late summer.

References

Acar, B. Beaman, M., & Porter, R.F. (1977). Status and migration of Birds of Prey in Turkey. in *World Conference on birds of prey*, R.D. Chancellor (ed.), pp. 182–7. Vienna, I.C.B.P.
Alexander, G., Mann, T., Mulhearn, C.J., Rowley, I.C.R., Williams, D. & Winn, D. (1967). Activities of foxes and crows in a flock of lambing ewes. *Australian Journal of Experimental Agriculture and Animal Husbandry*, 7: 329–36.
Alston, C.H. (1912). *Wild life in the west Highlands*. Glasgow.
Anon. (1846). Notes on a tour (by a Free Church Minister) in Shetland and Orkney. Sept 1845.
Armstrong, E.A. (1958). *The folklore of birds*. London, Collins.
Arnold, E.L. & Maclaren, P.I.R. (1938). Notes on the habits and distribution of the White-tailed Eagle in Iceland. *British Birds*, 34: 4–10.
Bannerman, D.A. (1956). *The birds of the British Isles*, vol 5. Edinburgh, Oliver & Boyd.
Banzhaf, W. (1937). Der Seeadler. *Dohrniana*, 16: 3–41.
Bauer, K. (1977). Present status of birds of prey in Austria. in *World conference on birds of prey*, R.D. Chancellor, (ed.) pp. 83–7. Vienna, I.C.B.P.
Baxter, E.V. & Rintoul, L.J. (1953). *The birds of Scotland*, vol 1. Edinburgh, Oliver & Boyd.
Bécsy, L. & Keve, A. (1977). The protection and status of birds of prey in Hungary. in *World conference on birds of prey*, R.D. Chancellor (ed.), pp. 125–9. Vienna, I.C.B.P.
Bent, A.C. (1961) *Life histories of North American birds of prey*. London, Dover.
Berg, B. (1923). *De sista örnarna*. Stockholm, Norstedt er Soener.
Berg, W., Johnels, A., Stöstrand, B. & Westermark, T. (1966). Mercury content in feathers of Swedish birds from the past 100 years. *Oikos*, 17: 71–85.
Bergman, G. (1961). The food of birds of prey and owls in Fenno-Scanida. *British Birds*, 54: 307–20.
Bergman, G. (1977). Birds of prey: the situation in Finland. in *World conference on birds of prey*, R.D. Chancellor (ed.), pp. 96–103. Vienna, I.C.B.P.
Beveridge, F.S. (1913). Sea Eagles in Argyll. *Scottish Naturalist*, No. 21: 190.
Bewick, T. (1821). *A history of British birds*, vol 1. Newcastle.
Bijleveld, M.F.I.J. (1974). *Birds of prey in Europe*. London, Macmillan.

Blackburn Mrs H. (1895). *Birds from Moidart and elsewhere*, Edinburgh, David Douglas.

Bogucki, Z. (1977). Status of the White-tailed Eagle in Poland. in *Report of WWF symposium on the White-tailed Eagle. Sept 1976*, pp. 31–2. Svanøy, Norway, WWF.

Bolam, G. (1912). *Birds of Northumberland and the eastern borders*. Alnwick, Blair.

Booth, E.T. (1881–87). *Rough notes on the birds observed during twenty-five years shooting and collecting in the British Isles*, vol 1. London.

Borg, K. Erne, K. Hanko, E. & Wanntorp, H. (1970). Experimental secondary methyl mercury poisoning in the Goshawk (*Accipiter g. gentilis* L.). *Environmental Pollution*, **1**: 91–104.

Borg, K., Wanntorp, H., Erne, K. & Hanko, E. (1965). *Kvicksilverförgiftningar bland vilt i Sverige*. Stockholm, State Veterinary Medical Institute (in Swedish).

Brodkorb, P. (1955). Number of feathers and weights of various systems in a Bald Eagle, *Wilson Bulletin*, **67**: 142.

Brodkorb, P. (1964). Catalogue of fossil birds, Part 2. *Bulletin of the Florida State Museum of Biological Sciences*, **8**; 195–335.

Brody, H. (1974). *Inishkillane*. London, Penguin.

Broley, C.L. (1947). Migration and nesting of Florida Bald Eagles. *Wilson Bulletin*, **59**: 3–20.

Broo, B. (1978). Project Eagle Owl, South West. in *Birds of prey management techniques*, T.A. Geer, (ed.), pp. 104–20. Oxford, British Falconers' Club.

Brown, L.H. (1976a). *British birds of prey*. London, Collins.

Brown, L.H. (1976b). *Eagles of the world*. Newton Abbot, David & Charles.

Brown, L.H. (1980). *The African Fish Eagle*. Folkestone, Bailey Bros & Swinfen Ltd.

Brown, L.H. & Amadon, D. (1968). *Eagles, hawks and falcons of the world*, vols 1 & 2. London, Country Life.

Bryden, J. & Houston, G. (1976). *Agrarian change in the Scottish Highlands*. Inverness, Martin Robertson & HIDB.

Buckley, T.E. & Harvie-Brown, J.A. (1891). *A vertebrate fauna of the Orkney Islands*. Edinburgh, David Douglas.

Cade, T. & Temple, S.A. (1977). The Cornell University Falcon Programme. in *World conference on birds of prey*, ed. R.D. Chancellor (ed.), pp. 353–69. Vienna, I.C.B.P.

Christensen, J. (1979). Den grønlandske Havørns *Haliaeetus albicilla groenlandicus* Brehms. ynglebiotop, redeplacering og rede. [The breeding habitat, nest-site and nest of the Greenland White-tailed Eagle] (in Danish, with English summary) *Dansk ornithologisk Forenings Tidsskrift*, **73**: 131–56.

Colquhoun, J. (1888). *The moor and the loch*. Edinburgh & London.

Conder, P. (1977). Legal status of birds of prey and owls in Europe. in *World conference on birds of prey*, R.D. Chancellor (ed.), pp. 189–93. Vienna, I.C.B.P.

Corkhill, P. (1980). Golden Eagles on Rhum. *Scottish Birds*, **11**: 33–43.

Cowles, G.S. (1969). Alleged skeleton of Osprey attached to carp. *British Birds*, **62**: 541–4.

Craighead, J.J. & Craighead, F.C. (1969). *Hawks, owls and wildlife*. New York, Dover.

Cramp, S. & Simmons, K.E.L. (1980). *Handbook of birds of Europe, the Middle East and North Africa*, vol 2. Oxford, Oxford University Press.

Davidson, Hilda R.E. (1964). *Gods and myths of northern Europe*. Harmondsworth, Middlesex, Penguin.

Davidson, Hilda R.E. (1969). *Scandanavian mythology*. London, Hamlyn.

Dementiev, G.P. & Gladkov, H.A. (1951). *The birds of the Soviet Union*. Jerusalem, Israel Programme for Scientific Translations, 1966.

Dennis, R.H. (1968). Sea Eagles. *FIBO Report*, No. 21: 17–21.

Dennis, R.H. (1969). Sea Eagles. *FIBO Report*, No. 22: 23–9.
Dennis, R.H. (1970). The oiling of large raptors by Fulmars. *Scottish Birds*, **6**: 198–9.
Deppe, H.J. (1972). Einige Verhaltensbeobach tungen in einem Doppelhorst von Seeadler (*Haliaeetus albicilla*) and Wanderfalke (*Falco peregrinus*) in Mecklenburg. *Journal of Ornithology*, **113**: 440–4. (in German).
Dick, Rev. C.H. (1916), *Highways and byways in Galloway and Carrick*. London, Macmillan.
Dixon, C. (1900). *Among the birds in northern Shires*. London.
Dixon, J.H. (1886). *Gairloch and guide to Loch Maree*. Edinburgh, Co-op. Printing Co.
Dornbusch, M. (1977). Der Seeadler *Haliaeetus albicilla* (L. 1758) in der Deutschen Demokratischen Republik. in *Report of WWF symposium of the White-tailed Eagle*. Sept 1976. pp. 17–18. Svanøy, Norway, WWF.
Dresser, H.E. (1871–81). *A history of the birds of Europe*. London.
Drosier, R. (1831). Account of an ornithological visit to the islands of Shetland and Orkney in the summer of 1828. *Magazine of Natural History*, **4**: 193–9.
Dunstan, T.C. (1978). Where Bald Eagles still soar. *National Geographic*, **153**: 186–199.
D'Urban, W.S.M. & Matthew, Rev. M.A. (1892). *The birds of Devon*. London, Porter.
Dyck, J., Eskildsen, J. & Møller H.S. (1977). The status of breeding birds of prey in Denmark 1975. in *World conference on birds of prey*, R.D. Chancellor (ed.), pp. 91–6. Vienna, I.C.B.P.
Eagle Clarke, W. (1912). *Studies in bird migration*, vol 2. Edinburgh.
Errington, P. (1946). Predation and vertebrate populations. *Quarterly Review of Biology*, **21**: 144–77, 221–45.
Evans, A.H. & Buckley, T.E. (1899). *A vertebrate fauna of the Shetland Islands*. Edinburgh, David Douglas.
Everett, M.J. (1971). The Golden Eagle survey in Scotland in 1964–68. *British Birds*, **64**: 49–56.
Feilden, H.W. (1872). The birds of the Faeroe Islands. *Zoologist*, **7**: 3210–25.
Fentzloff, C. (1977). Successful breeding and adoption of Sea Eagles (*Haliaeetus albicilla*). in *Papers on the veterinary medicine and domestic breeding of birds of prey*, J.E. Cooper & R.E. Kenward (eds), pp. 71–91. Oxford, British Falconers' Club.
Fentzloff, C. (1978). The breeding and release of the White-tailed Sea Eagle (*Haliaeetus albicilla*) as performed by the Deutsche Greifenwarte, Burg Guttenburg. In *Birds of prey management techniques*, T.A. Geer (ed.), pp. 97–103. Oxford, British Falconers' Club.
Fergusson, C. (1885). The Gaelic names of birds, Part 1. *Transactions of the Gaelic Society of Inverness*, **11**: 240–60.
Fevold, H.R. & Craighead, J.J. (1958). Food requirements of the Golden Eagle. *Auk*, **75**: 312–17.
Fiedler, W. (1970). Breeding of the White-tailed Sea Eagle *Haliaeetus albicilla* at Vienna Zoo. *International Zoo Yearbook*, **10**: 17–19.
Fiedler, W. (1977). On the captive breeding of the White-tailed Sea Eagle (*Haliaeetus albicilla*) and Griffon Vulture (*Gyps fulvus*) in Schönbrunn Zoo. in *World conference on birds of prey*, R.D. Chancellor (ed.), pp. 372–5. Vienna, I.C.B.P.
Fintha, I. (1976). The White-tailed Eagle (*Haliaeetus albicilla*) in Hortobágy. *Aquila*, **83**: 243–59.
Fischer, W. (1970). *Die Seeadler*. Wittenberg-Lutherstadt, A. Ziemsen Verlag.
Fisher, J. (1966a). *The Shell Bird Book*. London, Edbury Press & Michael Joseph.
Fisher, J. (1966b). The Fulmar population of Britain and Ireland 1959. *Bird Study*, **13**: 5–76.
Flerov, A.I. (1970). [On the ecology of the White-tailed Eagle in the Kandalaksha Bay.]

Proceedings of the Kandalaksha State Nature Reserve (Murmansk), No. VII: 215–32. (In Russian).

Forrest, H.E. (1907). *The vertebrate fauna of North Wales*. London, Witherby.

Forsman, D. (1981). Ruggnings förlopp hos och åldersbestamning av Havsörn (*Haliaeetus a. albicilla* L.). [Moult sequence and ageing in the White-tailed Eagle.] In *Projekt Havsörn i Finland och Sverige*, T. Stjernberg (ed.), pp. 164–94. (In Finnish with English summary) Helsinki, Jord-och Skogsbrukministeriet.

Friedman, H. (1950). The birds of North and Middle America. *US National Museum Bulletin*, No. 50 (6 vols): 793pp.

Galushin, V.M. (1977). Recent changes in the actual and legislative status of birds of prey in the USSR. in *World conference of birds of prey*, R.D. Chancellor (ed.), pp. 152–9. Vienna, I.C.B.P.

Gerrard, P. Naomi, Wiemeyer, S.N. & Gerrard, J.M. (1979). Some observations on the behaviour of captive Bald Eagles before and during incubation. *Raptor Research*, **13**: 57–64.

Geroudet, P. (1977). The reintroduction of the Bearded Vulture in the Alps. in *World conference on birds of prey*, R.D. Chancellor (ed.), pp. 392–437. Vienna, I.C.B.P.

Gladstone, H.S. (1910). *The birds of Dumfriess-shire*. London, Witherby.

Glutz von Blotzheim, U., Bauer, K. and Bezzel, E. (1971). *Handbuch der Vögel Mitteleuropas*, Band 3. Frankfurt, Akademische Verlagsgesellschaft.

Goldberg, E.D., Butler, P., Meier, P., Menzel, D., Paulik, G., Risebrough, R. & Stockel, L.F. (1971). Chlorinated hydrocarbons in the marine environment. [Report prepared by the panel on monitoring persistent pesticides in the marine environment of the Committee of Oceanography.] Washington, DC, National Academy of Science.

Gordon, S. (1955). *The Golden Eagle: King of birds*. London, Collins.

Gray, R. (1871). *The birds of the West of Scotland, including the Outer Hebrides*. Glasgow.

Grenquist, P. (1952). Förändringar i ejderns och svärtans förekomst i finska skärgården. *Riistatieteellisia Julkaisuja* [Papers on Game Research . No. 8. (In Swedish).

Gudmundsson, F. (1967). [The White-tailed Eagle.] in *The White-tailed Eagle*, B. Kjaran. pp. 97–134. Reykjavik, Bokfellsutgafan H.F. (In Icelandic).

Gunn, R.C. & Robertson, J.F. (1963). Lamb mortality in Scottish Hill flocks. *Animal Production*, **5**: 67–75.

Hagen, Y. (1976). Havørn og Kongeørn i Norge. *Viltrapport*, No. 1: 93pp. (In Norwegian with English summary).

Hancock, D. (1973). Captive propagation of Bald Eagles *Haliaeetus leucocephalus* – a review. *International Zoo Yearbook*, **13**: 244–9.

Hansen, K. (1979). Status over bestanden af Havørn *Haliaeetus albicilla groenlandicus* Brehm i Grønland i årene, 1972–74. [Population status for the Greenland White-tailed Eagle covering the years 1972–74.] *Dansk ornithologisk Forenings Tidsskrift*, **73**: 107–30. (In Danish with English summary).

Hario, M. (1981). Vinterutfodring av Havsörn i Finland. [Winter feeding of White-tailed Eagles in Finland.] in *Projekt Havsörn i Finland och Sverige*, T. Stjernberg (ed.), pp. 113–22. Helsinki, Jord-och Skogsbruksministeriet. (In Finnish with English summary).

Harper, J. (1876). *Rambles in Galloway*. Edinburgh, Edmonston & Douglas.

Harrison, C.J.P. (1980). A re-examination of British Devensian and early Holocene bird bones in the British Museum (Natural History). *Journal of archaeological Science*, **7**: 53–68.

Harrison, C.J.O. & Walker, C.A. (1973). An undescribed extinct fish-eagle from the Chatham Islands. *Ibis*, **115**: 274–7.

Harrison, C.J.O. & Walker, C.A. (1977). A re-examination of the fossil birds from the Upper Pleistocene in the London Basin. *The London Naturalist*, **56**: 6–9.

Harvie-Brown, J.A. (1906). *A fauna of the Tay Basin and Strathmore*. Edinburgh, David Douglas.

Harvie-Brown, J.A. (1902). On the avifauna of the Outer Hebrides. *Annals of Scottish Natural History*, No. 44: 201–2.

Harvie-Brown, J.A. (1904). An old Inverness-shire Vermin List. *Annals of Scottish Natural History*, No. 51: 185–6.

Harvie-Brown, J.A. & Buckley, T.E. (1887). *A vertebrate fauna of Sutherland, Caithness and West Cromarty*. Edinburgh, David Douglas.

Harvie-Brown, J.A. & Buckley, T.E. (1888). *A vertebrate fauna of the Outer Hebrides*. Edinburgh, David Douglas.

Harvie-Brown, J.A. & Buckley, T.E. (1892). *A vertebrate fauna of Argyll and the Inner Hebrides*. Edinburgh, David Douglas.

Harvie-Brown, J.A. & Buckley, T.E. (1895). *A vertebrate fauna of the Moray Basin*, 2 vols. Edinburgh, David Douglas.

Harvie-Brown, J.A. & Macpherson, Rev. H.A. (1904). *A vertebrate fauna of the North-west Highlands and Skye*. Edinburgh, David Douglas.

Helander, B. (1975). *Havsörnen i Sverige*. Bokusläningen, Uddevalla, Sweden, Svenska Naturskyddsföreningen. (In Swedish with English summary).

Helander, B. (1976). The White-tailed Sea Eagle – 1976 Swedish Report. in *Report of WWF Symposium on the White-tailed Sea Eagle*, Svanøy, Norway, WWF.

Helander, B. (1977). The White-tailed Sea Eagle in Sweden. in *World conference on birds of prey*, R.D. Chancellor (ed.), pp. 319–29. Vienna, I.C.B.P.

Helander, B. (1978a). White-tailed Sea Eagle, Annual status report 1977. in *World Wildlife Yearbook 1977–78*, P. Jackson (ed.), pp. 40–1. Morges, Switzerland, WWF.

Helander, B. (1978b). Feeding White-tailed Sea Eagles in Sweden. in *Endangered birds: techniques for preserving threatened species*, S.A. Temple (ed.), Madison, Wisconsin, University of Wisconsin Press.

Helander, B. (1979). White-tailed Sea Eagle, Annual status report 1978. in *World Wildlife Yearbook 1978–79*, P. Jackson, (ed.), pp. 22–3. Morges, Switzerland.

Helander, B. (1980). Färgringmärkning av havsörnen lägesrapport. [Colour banding of White-tailed Sea Eagles in northern Europe – a progress report.] *Fauna och flora*, **75**: 183–7. (In Swedish with English summary).

Helander, B. (1981a). Projekt Havsörn i Sverige. [Project Sea Eagle in Sweden.] in *Projekt Havsörn i Finland och Sverige*, T. Stjernberg (ed.), pp. 15–30. Helsinki, Jord-och Skogsbruksministeriet (In Swedish with English summary).

Helander, B. (1981b). Utfodring av Havsörn i Sverige. [Feeding White-tailed Eagles in Sweden.] in *Projekt Havsörn i Finland och Sverige*, T. Stjernberg (ed.), pp. 91–112. Helsinki, Jord-och Skogsbruksministeriet. (In Swedish with English summary).

Helander, B. (1981c). Nestling measurements and weights from two White-tailed Eagle populations in Sweden. *Bird Study*, **28**: 235–41.

Henderson, Isabel (1967). *The Picts*. London, Thames & Hudson.

Henriksson, K., Kappanen, E. & Helminen, M. (1966). High residues of mercury in Finnish White-tailed eagles. *Ornis Fennica*, **43**: 38–45.

Herrick, F.B. (1924). Life history of the Bald Eagle. *Auk*, **41**: 89–105, 213–31, 389–422 & 517–41.

Houston, D. (1977). The effect of Hooded Crows on hill sheep farming in Argyll, Scotland: Hooded Crow damage to hill sheep. *Journal of Applied Ecology*, **14**: 17–19.

Hudson, W.H. (1906). *British birds*. London.

Hunter, J. (1976). *The making of the crofting community*. Edinburgh, John Donald.

Hurrell, L.H. (1977). Breeding in skylight and seclusion facilities. in *Papers on the veterinary medicine and domestic breeding of diurnal birds of prey*, J.E. Cooper & R.E. Kenward (eds.), pp. 30–6. Oxford, British Falconers' Club.

Ingolfsson, A. (1961). The distribution and breeding ecology of the White-tailed Eagle *Haliaeetus albicilla* (L) in Iceland. Aberdeen, Unpublished Hons B.Sc. thesis.

Jensen, S., Johnels, A.G., Olsson, M. & Otterlind, G. (1969). DDT and PCB in marine animals from Swedish waters, *Nature*, 224: 247–50.

Jensen, S., Johnels, A.G., Olsson, M. & Westermark, T. (1972). The avifauna of Sweden as indicators of environmental contamination with mercury and chlorinated hydrocarbons. *Proceedings of the International Ornithological Congress*, 15: 455–65.

Jollie, M. (1947). Plumage changes in the Golden Eagle. *Auk*, 64: 549–76.

Jourdain, F.C.R. (1912). Extermination of the Sea Eagle in Ireland. *British Birds*, 5: 138–9.

Joutsamo, E. & Koivusaari, J. (1977). White-tailed Eagle in Finland 1970–76. in *Report from WWF symposium on the White-tailed Eagle. Sept 1976*. pp. 12–16. Svanøy, Norway, WWF.

Kampp, K. & Wille, F. (1979). Fødevaner hos den Grønlandske Havørn *Haliaeetus albicilla groenlandicus* Brehm. [Food habits of the Greenland White-tailed Eagle.] *Dansk ornithologisk Forenings Tidsskrift*, 73: 157–64. (In Danish with English summary).

Kirkwood, J.K. (1980). Energy and prey requirements of the young free-flying Kestrel. *Hawk Trust annual Report*, 1980: 12–14.

Koeman, J.H., de Goeij, J.J.M., Garssen-Hoekstra, J. & Pels, E. (1971). Poisoning of predatory birds by methyl-mercury compounds. *Meded. Rijksfac. Landbouwwe tensch. Gent.*, 36: 43–9.

Koeman, J.H., Hadderingh, R.H. & Bijleveld, M.F.I. (1972). Persistent pollutants in the White-tailed Eagle (*Haliaeetus albicilla*) in the Federal Republic of Germany. *Biological Conservation*, 4: 373–7.

Leigh, Miss M.M. (1928). The crofting problem 1780–1883. *Scottish Journal of Agriculture*, 11: 55pp.

Lewis, E. (1938). *In search of the Gyr Falcon*. London, Constable.

Lilford, Lord (1885). *Coloured figures of the birds of the British Islands*, vol 1. London.

Lockie, J.D. (1964). The breeding density of the Golden Eagle and Fox in relation to food supply in Wester Ross, Scotland. *Scottish Naturalist*, 71: 67–77.

Lockie, J.D. & Stephen, D. (1959). Eagles, lambs and land management on Lewis. *Journal of Animal Ecology*, 28: 43–50.

Lodge, G.E. (1946). *Memoirs of an artist naturalist*. London, Gurney & Jackson.

Love, J.A. (1977). The reintroduction of the Sea Eagle to the Isle of Rhum. *Hawk Trust annual Report*, 1977: 16–18.

Love, J.A. (1979). The daily food intake of captive White-tailed Eagles. *Bird Study*, 26: 64–6.

Love, J.A. (1980a). Birdwatching on Rhum. *Scottish Birds*, 11: 48–51.

Love, J.A. (1980b). White-tailed Eagle reintroduction on the Isle of Rhum. *Scottish Birds*, 11: 65–73.

Love, J.A. (1982). Harvie-Brown – a profile. *Scottish Birds*, 12: 49–53.

Low, Rev. G. (1813). *Fauna Orcadensis*. Edinburgh.

Low, Rev. G. (1879). *A tour through the Islands of Orkney and Shetland in 1774*. Edinburgh. (1978 reprint: Inverness, Melven).

Macaulay, Rev. K. (1764). *History of St Kilda.* Reprinted by James Thin, Edinburgh, 1974.
Macgillivray, W. (1886). *Descriptions of the rapacious birds of Great Britain.* Edinburgh.
Mackenzie, O.H. (1928). *A hundred years in the Highlands.* London, Geoffrey Bles.
MacNally, L. (1977). *The ways of an Eagle.* London, Collins.
Macpherson, Rev. H.A. (1892). *A vertebrate fauna of Lakeland.* Edinburgh, David Douglas.
Maestrelli, J.R. & Wiemeyer, S.N. (1975). Breeding Bald Eagles in captivity. *Wilson Bulletin,* No. 87: 45–53.
Marquiss, M. & Newton, I. (1982). The Goshawk in Britain. *British Birds,* **75**: 243–60.
Martin, M. (1716). *Description of the Western Isles of Scotland.* (1976 reprinted, Edinburgh, James Thin).
Maxwell, Sir H. (1907). *Memories of the month,* first series. London, Edward Arnold.
McIan, R.R. (1848). *Gaelic gatherings.* London, Ackerman.
McWilliam, J.M. (1936). *The birds of the Firth of Clyde.* London.
Meinertzhagen, R. (1930). *Nicoll's birds of Egypt,* vol 2. London.
Meinertzhagen, R. (1959). *Pirates and predators.* Edinburgh, Oliver & Boyd.
Meyburg, B.-U. (1977). Protective management of eagles by reduction of nestling mortality. in *World conference on birds of prey,* R.D. Chancellor (ed.), pp. 387–92. Vienna, I.C.B.P.
Meyburg, B.-U. (1978). Productivity manipulation in wild eagles. in *Birds of prey management techniques,* T.A. Geer (ed.), pp. 81–92. Oxford, British Falconers' Club.
Mitchell, W.R. & Robson, R.W. (1976). *Lakeland birds: a visitors' handbook.* Clapham, Dalesman Books.
Moore, N.W. (1955). The past and present status of the Buzzard in the British Isles. *British Birds,* **50**: 173–97.
More, A.G. (1865). Distribution of birds in Great Britain during the nesting season. *Ibis,* Ser. 2, **1**: 1–27.
Mori, S. & Nakagawa, H. (1981). The fauna of bird [sic] and its character in the Shiretoko Peninsula, Hokkaido, Japan. in *Report of the Survey on vertebrate communities in Shiretoko Peninsula, Hokkaido, Japan,* pp. 89–97. Sapporo, Section of Nature Conservation, Hokkaido Government, (in Japanese, English summary).
Morris, Rev. F.O. (1851). *A history of British birds,* vol 1. London.
Muirhead, G. (1889). *The birds of Berwickshire,* vol 1. Edinburgh, David Douglas.
Mundy, P.J. & Ledger, J.A. (1976). Griffon Vultures, carnivores and bones. *South African Journal of Science,* **72**: 106–10.
Murton, R.K. (1971). *Man and birds.* London, Collins.
Nelson, M.W. (1978). Preventing electrocution deaths and the use of nesting platforms on power lines. in *Birds of prey management techniques,* T.A. Geer (ed.), pp. 42–6. Oxford, British Falconers' Club.
New statistical account of Scotland. (1835–45). 15 vols Edinburgh.
Newton, I. (1972). Birds of prey in Scotland: some conservation problems. *Scottish Birds,* **7**: 5–23.
Newton, I. (1979). *Population ecology of raptors.* Berkhamsted, Poyser.
Niethammer, G. (1938). *Handbuch der deutschen Vogelkunde,* Band 2. Leipzig.
Norderhaug, M. (1977). Preliminary survey on recent data about the White-tailed Eagle *Haliaeetus albicilla* in Norway. in *Report on WWF symposium on White-tailed Eagle Sept 1976.* pp. 22–30. Svanøy, Norway, WWF.
Nye, P.E. (1981). A biological and economic review of the hacking process for the

restoration of Bald Eagles. (Presented at the International Bald Eagle/Osprey Symposium October 28–29, 1981. Montreal, Canada.) in *Federal aid to endangered species, New York project E-1, Performance report*, 16pp.

Nye, P.E. (1981). Restoring the Bald Eagle in New York. *The Conservationist*, **37**: 8–13.

O'Dell, A.C. & Walton, K. (1962). *The Highlands and Islands of Scotland*. Edinburgh, Nelson.

Oehme, G. (1961). Der Bestandsentwicklung des Seeadlers in Deutschland mit Untersuchungen zur Wahl der Brutbiotopen. In *Beiträge zur Kenntnis deutscher Vögel*, H. Schildmacher (ed.), pp. 1–61, Jena.

Oehme, G. (1969a). Schutz des Seeadlers. *Falke*, **16**: 54–60.

Oehme, G. (1969b). Population trends in the White-tailed Sea Eagle in North Germany. in *Peregrine Falcon populations*, J.K. Hickey (ed.), pp. 351–2. Madison, Wisconsin, University of Wisconsin Press.

Oehme, G. (1977). Seeadler, in *Die Vogelwelt Mecklenburgs*. G. Klafs & J. Stubbs (eds.), pp. 134–5. Jena, Fischer.

Old statistical account, see Sinclair (1791–99).

Olsson, V. (1972). Revir biotop och boplatsval hos svenska havsörnar *Haliaeetus albicilla*. *Vår Fågelvarld*, **31**: 89–95.

Opdam, P. & Muskens, G. (1976). Use of shed feathers in population studies of *Accipiter* hawks. *Beaufortia*, **24**: 55–62.

Paton, E.R. & Pike, O.G. (1929). *The birds of Ayrshire*. London, Witherby.

Pennant, T. (1774). *British zoology*. London.

Postupalski, S. & Holt, J.B. (1975). Adoption of nestlings by breeding Bald Eagles. *Raptor Research*, **9**: 18–20.

Puşcarin, V. & Filipaşcu, A. (1977). The situation of birds of prey in Rumania 1970–74. in *World conference on birds of prey*, R.D. Chancellor (ed.), pp. 148–52. Vienna, I.C.B.P.

Randla, T. & Oun, A. (1980). Kaljukotkas ja merikotkas Eestis 1970 – ndail aastail. [The Golden Eagle and the White-tailed Sea Eagle in Estonia in the 1970s.] *Eesti Loadus*, **23**: 510–15, 543.

Ratcliffe, D. (1980). *The Peregrine Falcon*. Berkhamsted, Poyser.

Richmond, K. (1959). *British birds of prey*. London, Lutterworth.

Ritchie, J. (1920). *The influence of man on animal life in Scotland*. Cambridge, CUP.

Rowley, I. (1970). Lamb predation in Australia: incidence, predisposing conditions and the identification of wounds. *CSIRO Wildlife Research*, **15**: 79–123.

Rudebeck, G. (1951). The choice of prey and modes of hunting of predatory birds with special reference to their selective effect. *Oikos*, **3**: 200–31.

Ruger, A. (1975). Seeadlerschutz in Schleswig-Holstein. *Wild und Hund*, **78**: 322–5.

Ruger, A. (1981). Bestandstützung durch Adoptionsverfahren Erfahrungen mit Seeadlern in Schleswig-Holstein. *Natur und Landschaft*, **56**: 133–5.

Salomonsen, F. (1950). *Grönlands Fugle*. Copenhagen, Munksgaard.

Salomonsen, F. (1967). *Fuglene på Grønland*. Copenhagen, Rhodos.

Sandeman, P. (1957a). The breeding success of Golden Eagles in the southern Grampians. *Scottish Naturalist*, **69**: 148–52.

Sandeman, P. (1957b). The rarer birds of prey; their status in the British Isles. *Pandion haliaetus*, *British Birds*, **50**: 129–55.

Sandeman, P. (1965). Attempted reintroduction of White-tailed Eagle to Scotland. *Scottish Birds*, **3**: 411–12.

Saunders, H. & Eagle Clarke, W. (1927). *Manual of British birds*. London, Gurney & Jackson.

Saurola, P. (1978). Artificial nest construction in Europe. in *Birds of prey management techniques*, T.A. Geer (ed.), pp. 72–80. Oxford, British Falconers' Club.
Saurola, P. (1981). Ringmärkning och aterfynd av Finländska Havsörnar. [Ringing of White-tailed Sea Eagles in Finland.] in *Projekt Havsörn i Finland och Sverige*, T. Stjernberg (ed.), pp. 135–45. Helsinki, Jord-och Skogsbruksministeriet (In Swedish with English summary).
Saxby, H.L. (1874). *The birds of Shetland*. Edinburgh, MacLachlan & Stewart.
Schiöler, E.L. (1931). *Danmarks Fugle*, vol 3. Copenhagen.
Schnurre, O. (1956). Ernä hrungsbiologische Studien an Raubvögeln und Euten der Darsshalbinsel (Mecklenberg). *Beiträge Vogelkunde*, 4: 211–45.
Seebohm, H. (1883). *A history of British birds*, vol 1. London.
Sharrock, J.T.R. (ed.) (1976). *The atlas of breeding birds in Britain and Ireland*. Berkhamstead, Poyser.
Sibley, C.G. & Ahlquist, J.E. (1972). A comparative study of the egg-white proteins of non-passerine birds. *Peabody Museum of Natural History, Yale University Bulletin*, 39: 1–276.
Sim, G. (1903). *The vertebrate fauna of Dee*. Aberdeen.
Sinclair, Sir J. (1791–99). (ed.) *Old statistical account of Scotland*. 21 vols Edinburgh.
Sládek, J. (1977). The status of birds of prey in Czechoslovakia. in *World conference on birds of prey*, R.D. Chancellor (ed.), pp. 87–91. Vienna, I.C.B.P.
Sloan, J.H. (1908). *Galloway*. London, A. & C. Black.
Soot-Ryen, T. (1941). Egg-og dunvaer i Troms fylke. *Tromsø Mus. Arsh*, No. 62.
Southern, W.E (1964). Additional observations on winter Bald Eagle populations: including remarks on biotelemetry techniques and immature plumages. *Auk*, 76: 121–37.
Sprunt, A., Robertson, W.B., Postupalsky, S., Hensel, R.H., Knoder, C.E. & Ligas, F.J. (1973). Comparative productivity of six bald eagle populations. *Transactions of the North American Wildlife and Natural Resources Conference*, 38: 96–106.
Stager, K. (1964). The role of olfaction in food location by the Turkey Vulture (*Cathartes aura*). *Los Angeles County Museum Contribution in Science*, No. 81.
Steele-Elliot, T. (1895). Observations on the fauna of St Kilda. *Zoologist*, 19: 281–6.
Stjernberg, T. (1981). Projekt Havsörn i Finland. in *Projekt Havsörn i Finland och Sverige*, T. Stjernberg (ed.), pp. 31–60. Helsinki, Jord-och Skogsbruksministeriet. (With English summary).
St. John, C. (1849). *A tour in Sutherland*. Edinburgh, David Douglas.
St. John, C. (1893). *Wild sports and natural history in the Highlands*. Edinburgh.
Suetens, W. & von Groenendel, P. (1968). Notes sur deux oiseaux de proie de la Yugoslavie orientale: Faucon sacre et Pygargue à queue blanche. *Gerfaut*, 58: 78–93.
Thacker, R. (1971). Estimations relative to birds of prey in captivity in the USA. *Raptor Research News*, 5: 108.
Tulloch, J.S. (1904). Albino Sea Eagle in Yell, Shetland. *Annals of Scottish Natural History*, No. 52: 245.
Tulloch, R. (1978). The eagle and the baby. *Scots Magazine*, 109: 260–4.
Ussher, R.J. & Warren, R. (1900). *The birds of Ireland*. London, Gurney & Jackson.
Uttendörfer, O. (1939). *Die Ernährung der Deutschen Raubvogel und Eulen*. Neudamm, Neumann.
Uttendörfer, O. (1952). *Neue Ergebnisse über die Ernährung der Greifvögel und Eulen*. Stuttgart, Eugen Ulmer.
Vagliano, C. (1977). The status of birds of prey in Greece. in *World conference on birds of prey*, R.D. Chancellor (ed.), pp. 118–25. Vienna, I.C.B.P.

Vaurie, C. (1965). *The birds of the Palaearctic fauna*. London, Witherby.
Venables, L.S.V. & Venables, U.M. (1955). *The birds and mammals of Shetland*. Edinburgh, Oliver & Boyd.
Wagner, F.H. (1972). Quoted in 'Eagles and sheep: a viewpoint', by E.G. Bolen, *Journal of Range Management*, **28**: 11–17.
Wallace, Rev. J. (1693). *A description of the Isles of Orkney*. Edinburgh.
Walls, G.L. (1942), *The vertebrate eye*, London.
Weir, D.N. (1973). A case of lamb-killing by Golden Eagles. *Scottish Birds*, **7**: 293–301.
Wille, F. (1977). Projekt Nagtoralik, in *Report from WWF symposium on the White-tailed Eagle. Sept 1976*, pp. 7–11. Svanøy, Norway, WWF.
Wille, F. (1979). Den grønlandske Havørns *Haliaeetus albicilla groenlandicus* Brehm. fødevalg – metode og foreløbige resultater. [Choice of food of the Greenland White-tailed Eagle – method and preliminary results.] *Dansk ornithologisk Forenings Tidsskrift*, **73**: 165–70.
Willgohs, J.F. (1961). *The White-tailed Eagle* Haliaeetus albicilla albicilla *(L.) in Norway*. Bergen, Arbok for Universitetet.
Willgohs, J.F. (1963). *Havørnen*. Bergen. Vestlandske Naturvenförening.
Willgohs, J.F. (1969). Status and prospects of Eagles in Norway. *Biological Conservation*, **2**: 71–2.
Williamson, K. (1970). *The Atlantic Islands*. London, Routledge & Kegan Paul.
Williamson, K. (1975). Birds and climatic change. *Bird Study*, **22**: 143–64.
Witherby, H.F., Jourdain, F.C.R., Ticehurst, N.F. & Tucker, B.W. (1943). *The handbook of British birds*, vol. 3. London, Witherby.
Wolley, J. (1902). *Ootheca Wolleyana*, vol. 1. A. Newton (ed.). London, R.H. Porter.
World Wildlife Fund (1976). [Manifesto on] 'Re-introductions: techniques and ethics'. In *Wildlife introductions to Great Britain*. Report by the working group on introductions, UK Committee for international nature conservation. London, Nature Conservancy Council.
Wormell, P. (1976). The Manx Shearwaters of Rhum. *Scottish Birds*, **9**: 103–18.
Yarrell, W. (1871). *A history of British birds*, vol. 1. London, Van Voorst.

Index

Page numbers in italic type indicate illustrations in the text

Accipiter gentilis, see Goshawk
Accipiter nisus, see Sparrowhawk
Accipitres,
  classification of, 5–13
African Fish Eagle (*Haliaeetus vocifer*), 8, 9, 81
  distribution of, 13
  plumage patterns on underside, 12
age,
  determination: of nestlings and wing length, 19; by plumage, 17–18, 19, 20, 21
  longevity, 45, 46, 78
  and mortality, 77
Albania, 28
albinos, 13–14, 45, 46
Algeria, 28
American Bald Eagle, see Bald Eagle
Aquila audax, see Wedge-tailed Eagle
Aquila chrysaetus, see Golden Eagle
Aquila heliaca, see Imperial Eagle
Aquila pomerina, see Lesser Spotted Eagle
Austria,
  Sea Eagle records, 29
  Vienna Zoo, 62, 73, 141

Bald Eagle (*Haliaeetus leucocephalus*), 2, 8, 10, 11
  courtship of, 57
  dispersion from breeding sites, 77
  eggs: incubation of, 65; transplantation of uncontaminated, 140
  fostering of eaglets, 147, 149
  geographical distribution, 11, 13; north to south, and size, 24
  nests of: artificial, and conservation, 134; Great Horned Owls nesting in, 54
  plumage: patterns on underside, 12; weight of, 14
  productivity of, 138
  release of, from captive pairs, 141
  restocking New York State with, 148–9, 150
  sexual dimorphism for size, 24
Bald Eagle, Alaska race, (*Haliaeetus leucocephalus alascanus*),
  restocking New York State with 149, 150
  sexual dimorphism, 24
  size of, 11
bathing by Sea Eagles, 181–2, 189, 192
BBC television,
  filming of Sea Eagles, 192
beak, 79, 80
  dimensions of, and sex determination, 19, 22, 23, 174
  measurement prior to release, 174, 175
Bearded Vulture (*Gyps fulvus*),
  reintroduction in Alps, 149
Brahminy Kite (*Haliastur indus*), 6, 7
breeding failure,
  causes of, 72, 136–8
brood sizes, 41, 70–2
  and nesting attempts, 71–2
Bulgaria, 26, 28
Burrian eagle, 109, 110
Buzzard (*Bursarellis nigricollis*), 8

buzzards, 90
  classification of, 5, 6
  nests of, used by Sea Eagles, 54
  pellet content and diet, 84
  skull of, *80*
  talon-grappling, by 55
  (*see also* Honey Buzzard)
calls, 4
  chick food-begging, 4; tape record of, and feeding captive chick, 145
  courtship, 4, 56, 57
  development of, 68
  sexual differences in, 18
Canary Islands,
  Sea Eagle records, 28
Canna, 193–4
  Sea Eagle decline, 41, 115
Catharacta skua, *see* Great Skua
Cathartes aura, *see* New World Vulture
chicks, *see* eaglets; fostering
China,
  occurrence of Sea Eagle, 26, 30
Circus aeruginosus, *see* Marsh Harrier
classification, 2
  of Falconiformes, 5–13
combat,
  between eagles, 55; talon-grappling, 55, 56
conservation,
  artificial eyries and breeding density, 134
  and forestry practices, 133
  by legislation, 132–3
  raising chicks in captivity, 141–4
  rescue of eggs from wild, 145
  by restocking, 148–9, 152
  reward schemes for, 133
  transplantation of uncontaminated eggs, 140
  uncontaminated food provision, 139, *140*
  (*see also* fostering)
Cormorant (*Phalacrocorax carbo*),
  colony in northern Norway, 165
Corsica, 27
courtship, 55–8, 59, 206
  calls, 4, 56, 57
  coition, 57, 58, 203
  display flights, 56–7
  female soliciting, 58, 59
  talon-grappling, 55, 56, 56, 57
Cyprus,
  Sea Eagle records, 28
Czechoslovakia,
  occurrence of Sea Eagles, 26, 28

decline of Sea Eagles, 115–16, 153–4 and
  accessibility of nesting sites, 114, 115, 122, 124
  and climate, 116–17
  competition from other species, 117

and food supply, 117
  fragmentation of habitat, 111, 131
  and Fulmars, 93, 117–8, 160
  and human persecution, 118–30
  and human population changes, 118–19
  and sheep farming, 118–19
  (*see also* destruction of Sea Eagles; eggs, collection of; toxic chemicals)
deer forests,
  and survival of Golden Eagles, 125, 128, 129
Demark, 26–7, 31
  protective legislation, 132
destruction of Sea Eagles, x–xi, 108, 111
  extinction in Britain, 45, 46, 115–16
  generation of campaign for, 101
  by pesticides, 30, 31, 71–2, 202
  by poison, 77; in carrion, 27, 28, 120, 125, 132, 201
  rewards for, 107–8, 111, 112–13, 114, 121
  by shooting, 27, 39, 41, 42, 44, 45, 77, 111, 119–20, 121, 131
  by trapping, 42, 113–14, *114*
  (*see also* decline of Sea Eagles; eggs, collection of; toxic chemicals)
diet, 86–8
  in captivity, 171–2; amount consumed, 172
  carrion in, 87, 88, 104, 105, 125, 201; from domestic stock, 101; from Red Deer, 197
  of eagle species, 8
  eggs of bird species, 93–4
  and pellet composition, 84, 86, 165
  seasonal variation in, 87, 88, 89
  (*see also* feeding; hunting)
dispersion from breeding sites, 74–7
  and return, 77
displays, *see under* flight
distribution, 8, 11, 13, 25–46
  fragmentation by habitat loss, 111, 114, 131
  of Greenland race, 25–6
  maps of, 13, 26; Scotland, 127, 129
  north to south, and size, 24, 63, 64
  in Scotland, 125–6, 127, 129

eaglets, 67, 69, 167
  age of: and wing length, 19; and first flight, 70
  brooding of, 65, 68
  calcium requirement, 86
  collection of, from Norway, 191, 192, 193, 164–6
  development of, 68–70, *69–70*
  feeding of, in the wild, 65–6, 66, 67, 68, 69, 87; and calls, 4

genetic variability of reintroduced stock, 166
raised in captivity, 141, 143–4, 145, 167–72; diet and amount consumed, 171–2; feeding of, 167, *168*, (and tape record of food-begging calls) 145; illnesses of, 169–70
sibling aggression, 72–3
weights of, 66
(*see also* fostering)
Eekholt Wildlife Park, 62–3, 143–4
eggs, 8
and breeding failure, 72
and brood sizes, 70–2
characteristics of: colour, 63; shape, 63; size, 63, 64; weight, 63
clutch sizes, 59, 61–3, *61*; four-egg, 61–2
collection of, 123–4; in England 111; in Hebrides, 42, 43; in Orkney, 44; in Rum, 41; in Scotland, 38, 39, 40, 121, 122; in Shetland, 45
hatching of, 65
incubation of, 63–5, *64*; duration of, 65; and food provision by male, 65, 66
laid in captivity, 62–3, 142
laying dates and latitude, 58, 60
legislation for protection of, 26
pesticides: and infertility of, 136; and shell thickness, 136, 137
rescue from wild, 145
serial hatching of, and chick development, 147
sizes of, and geographic distribution, 63–4
transplantation of uncontaminated, 140
Egypt, 27
Egyptian Vulture (*Neophron percnopterus*), 6, 7
Eider,
hunting of, 94, 95, 190
increase in Britain, 117
Eigg,
decline of Sea Eagles, 13, 41, 115
England,
decline and destruction of Sea Eagles, 111, 121
dispersion from, 77
records of Sea Eagles, 34–5; map, 33; prehistoric, 108, 109
extinction of Sea Eagles, 27, 28–9, 32
in Britain, 45, *46*, 115–16
eyes of Sea Eagles, 81
eyries, *see* nests

Fair Isle, 44, 115, 198
reintroduction attempt, 158–61; and Fulmars, 93, 118, 160
Falco cherrug, *see* Saker
Falco columbarius, *see* Merlin
Falco peregrinus, *see*, Peregrine
Falco subbuteo, *see* Hobby
Falco tinnunculus, *see* Kestrel
Falconiformes,
classification of, 5–13
evolution in, 5
falconry, ix, x
techniques of, used in reintroduction, 167, 169, *170*; hacking 148, 149, *150*, 156, 196; *see also* tethering
Faroe, 98
Nevtollur (tax system) and eagle destruction, 107
Sea Eagle records, 27
feeding, 74, *140*
of chicks, 65–6, 66, 67, 68, 69, 87; and calls, 4
and food intake, 82–3, 84; rate of, 84
frequency of, 81; and crop capacity, 84
released Sea Eagles, 159, 178, *179*, 181, 191; behaviour during, 186–7, *188*, 192
(*see also* diet; hunting)
Finland, 31
conservation: artificial eyries and breeding density, 134; protection of eyries, 133; protective legislation, 132; uncontaminated food provision, 139
dispersion of Sea Eagles from, 74, 75, 76
nesting sites, 48
prey taken: birds, 87, 94–5, 96–7; fishes, 87, 92; mammals, 87, 99
productivity of Sea Eagles, 138
toxic chemicals and breeding failure, 136
Fish Eagles, 6, 7, 56
courtship of, 56
fledging,
time of, and latitude, 73–4
flight,
display, 159, 203, 206; and calls, 4
duration of, from radio tracking, 81, 185
eating prey in, 187, *188*
first, and eaglet age, 70
height of, 160
identification during: by plumage patterns, *12*; by silhouettes, *19*, *20*, *21*
on release, 176, *177*, 178
talon-grappling during, 55, 56, 57, 159, 191, 203, 206
when hunting, 81
fossil records, 8, 34
fostering,
in captivity, 145
the runt of a brood, 73, 146–7
to wild pairs, 144–7
Foula,
Bonxies on, 44
records of Sea Eagles, 44

222   Index

France,
  Sea Eagle record, 27
Fulmar,
  caught by Sea Eagles, 93, 160
  increase in Britain, 117–18
  oil of, and death of predatory birds, 93,
    117–18, 160; on Rum, 161

Germany, 26, 29
  conservation: fostering to wild pairs,
    144–6; protection of eyries, 133, 134,
    135; protective legislation, 132
  destruction of Sea Eagles, 108, 113
  diet, 87; bird species taken, 93, 96; fish
    species taken, 91–2; seasonal variation
    in, 88, 89
  dispersion of Sea Eagles from, 74–7
  Eekholt Wildlife Park, 62–3, 143–4
  egg-laying dates, 58, 60
  nesting sites, 47, 48; density of, 49
  toxic chemicals and breeding failure,
    137, 138
Golden Eagle (*Aquila chrysaetos*), 2, 126
  aggression towards Sea Eagles, 205
  classification of, 6
  destruction of, 121
  distribution of, 38, 39, 41, 43, 125–6;
    maps, 127, 129; on Rum, 163, 202
  eggs: clutch size, 62; collection of, 123–4;
    laying period, 59
  in falconry, ix, x
  food consumption by, 83
  habitats of, 125, 127, 128, 129, 205;
    nesting sites, 126
  identification in the field, 3
  and lambs, 102, 103
  longevity of, 78
  moulting of, 14–15, 17
  skull of, *80*
  talon-grappling by, 55
  toxic chemicals and breeding difficulties,
    202
Goshawk (*Accipiter gentilis*),
  decline of, 128
  in falconry, ix
  pellet content and diet, 84
  prey to Sea Eagle, 92
  recolonisation by, 151
Great Bustard (*Otis tarda*), 149
Great Skua (*Catharacta skua*), 44
Greece, 26, 28
  bandit in nest of Sea Eagle, 52
  egg-laying period, 58, 60
  protective legislation, 132
Greenland,
  brood sizes, 72
  egg-laying period, 60
  diet, 86, 87; level of fish in, 202; prey
    taken, 90, 95, 100

nests: linings of, 52; sites of, 49, 124
  protective legislation, 27, 132
  (*see also* White-tailed Sea Eagle,
    Greenland race)
Grey-headed fishing eagle (*Ichthyophaga
  ichthyaetus*), 6, 7
Gypohierax angolensis, *see* Vulturine Fish
  Eagle

hacking,
  Bald Eagles, 149, *150*
  Peregrines, 148
  Sea Eagles, 156, 196
  (*see also* falconry; tethering)
Haliaeetus spp.,
  classification of, 6, 8–13
  courtship displays by, 55–7
Haliaeetus leucocephalus, *see* Bald Eagle
Haliaeetus leucocephalus alascanus, *see* Bald
  Eagle, Alaska race
Haliaeetus leucogaster, *see* White-bellied Sea
  Eagle
Haliaeetus leucoryphus, *see* Pallas' Sea Eagle
Haliaeetus pelagicus, *see* Steller's Sea Eagle
Haliaeetus piscator, 8
Haliaeetus sanfordi, *see* Sanford's Sea Eagle
Haliaeetus vocifer, *see* African Fish Eagle
Haliaeetus vociferoides, *see* Madagascar Fish
  Eagle
Haliastur indus, *see* Brahminy Kite
harriers,
  classification of, 5, 6
  (*see also* Hen Harrier, Marsh Harrier)
Harris, 193, 204
  decline of Sea Eagles, 43, 115
  hunting by Sea Eagles, 100
Hebrides,
  Golden Eagles on, 125–6
  Sea Eagles: brood sizes of, 71; clutch
    sizes of, 59, 61; decline of, 115; egg-
    laying dates, 58–9; lambs and, 102–3;
    nesting site records, 41–3; protection
    of, 120–1
  (*see also individual islands*)
Hen Harrier, (*Circus cyaneus*),
  talon-grappling by, 55
Hobby (*Falco subbuteo*), 128
Honey buzzard (*Pernis apivorus*, 128)
Hungary,
  pesticides and destruction of Sea Eagles,
    137
  use of eyries by bird species, 54
hunting, 79, 81
  for birds, 92–6, *95*; dealing with large
    specimens, 94; debilitated prey, 93;
    species taken, 92–3, 94–6
  for fish, 79, 88–92, *89*, *92*; adaptations
    for, 79, *80*, 81; behind boats, 90;
    dealing with large specimens, 91, *92*;

Index

piracy, 88, 90; species taken, 90–1, 91–2
 for mammals, 99–106; domestic stock, 101–6; species taken, 99–100
 other prey, 96, 97; human young, 97–8, 99
 by reintroduced Sea Eagles, 160, 201
 (*see also* diet; feeding)

Iceland, 26, 27, 75
 bird species taken by Sea Eagles, 95
 brood size, 71, 72
 chick sibling aggression, 73
 conservation: protection of eyries, 134; protective legislation, 132
 egg-laying period, 60
 nesting sites, 49; accessibility of, 124
Ichthyophaga australis, 8
Ichthyophaga ichthyaetus, *see* Grey-headed Fishing Eagle
Ichthyophaga nana, *see* Lesser Fishing Eagle
identification in the field,
 by plumage, 12, 13–14, 17–18, 186
 by silhouettes, 18, 19, 20, 21
 by wing tags, 194, 195
 (*see also* ringing)
immatures,
 appearance of, 4, 189
 destruction of, 77, 121–2
 dispersion of, 74–7; and return, 77
 feeding grounds of, 74
 pellet regurgitation, 85
 recolonisation by, in Scotland, 203–4
Imperial Eagle (*Aquila heliaca*), 54, 147
imprinting,
 and reintroduction by fostering, 156
Institute of Terrestrial Ecology (ITE), xi, 163
introduction,
 of species: definition of, 152; risks to ecosystem, 152
Iran, 26, 28
Iraq, 26, 28
 egg-laying period, 58, 60
Ireland,
 clutch sizes, 59
 decline and destruction, 115, 118, 120
 distribution of Sea Eagles, 32–3
 nests: construction of, 52; sites of, 49
 records of Sea Eagles, 112; prehistoric, 109
Isle of Man,
 Sea Eagle records, 33, 34, 111; last, 114
Israel, 28
 egg-laying period, 58, 60
 destruction of Sea Eagles, 137
 reintroduction intention, 143
 Tel-Aviv Zoo: breeding in captivity, 62, 96, 141–3, 142

Italy,
 protective legislation, 132
ITE, *see* Institute of Terrestrial Ecology

Japan, 26, 30
juveniles,
 appearance, 14, 17–18, 82, 179, 184
 mortality of, 77
 recolonisation by, in Scotland, 203–4
 silhouettes of, 18, 19
 tearing up prey, 85
 tethered and hooded, 173, 174; for measurements, 174, 175
 threatening behaviour, 167, 168

Kestrel (*Falco tinnunculus*),
 in falconry, ix
 food consumption by, 83
 talon-grappling by, 55
kites, 5, 6, 7, 54, 90
 (*see also* Red Kite; Brahminy Kite)
Korea, 26, 30

Lake District,
 Sea Eagles: destruction of, 111–12; lambs and, 102; records of, 111
Lebanon, 28
legislation,
 Sea Eagle protection, 27, 31, 132; and increase in raptors, 154; and reintroduction, 153, 154
Lesser Fishing Eagle (*Ichthyophaga nana*), 6, 7
Lesser Spotted Eagle (*Aquila pomerina*), 147
Lewis,
 Sea Eagles: dead sheep taken by, 101; decline of, 42; sighting of, 194
longevity, 45, 46, 78
Lundy Island,
 Sea Eagle records, 33, 34, 111

Madagascar Fish Eagle (*Haliaeetus vociferoides*), 8, 9
 distribution of, 13
 plumage patterns on underside, 12
Malta, 27
Manchuria, 30
Manx Shearwaters (*Puffinus puffinus*), on Rum, 161
Marsh Harrier (*Circus aeruginosus*),
 decline of, 128
 predator of Sea Eagle nests, 72
Merlin (*Falco columbarius*), 93
migration, *see* dispersion
Milvus milvus, *see* Red Kite
mobbing,
 of released Sea Eagles, 159; by gulls, 92; by Hooded Crows, 181, 189; when moulting, 160; by Peregrine, 159

Mongolia, 26, 30
moulting, *see under* plumage
Mull,
    Sea Eagle decline, 40–1, 115
myths and legends, 2–3, 37, 81, 101, 109, *110*
    baby-snatching by Sea Eagles, 97–8, *99*
    medicinal, 3, 107
    symbolic eagles, 2, 108–9

names,
    placenames with eagle derivations: in Faroe, 27; in Hebrides, 43; in Lake District, 111; in Scotland, 35, 36, 37, 41; in Shetland 44–5; in Skye, 42
    for Sea Eagle, 1–2
Nature Conservancy Council (NCC), xi, 166
    and reintroduction project, 161, 162, 163
Neophron percnopterus, see Egyptian Vulture
nest building,
    attempts at, and brood size, 71–2
    in captivity, *142*, 143, 144
    division of duties, 52
    last recorded dates of, 114–16
    materials used, 51–2, *53*
    month started, 54
    by reintroduced Sea Eagles, 203, 206
    speed of, 53
nesting sites, 47–52, *50*, *53*, 165
    accessibility of, 43, 51, 114, 115, 122, 124
    change of, 45, 51
    conservation of, 133–4, *135*
    densities of, 49
    long tenancy of, 37, 42, 45, 51, 111
    in nests of other species, 54
    picturesque situations of, 41, 51–2
    records of, 32–45, 126; map, 33
    of reintroduced Sea Eagles, 203
nests,
    predators of, 72
    robbing of, *39*; (*see also* eggs, collection of)
    sizes of, 52, *53*; bowl dimensions, 52
    use of, by other species, 54
New World Vulture (*Cathartes aura*), 81
Norway, 31–2
    accessibility of nesting sites, 124
    brood size, 71, 72
    collection of eaglets from, 191, 192, 193, 164–6
    destruction of Sea Eagles, 113–14; bounty schemes for, 107–8
    dispersion of Sea Eagles from, 74, 75, 76
    eggs: clutch size, 61, 62, 63; laying period, 58, 60
    fledging, 74
    hunting for fish, 88
    lambs and Sea Eagles, 103

Nature Reserves, 133
nests: construction of, 52; density of, 49, sites of, 47, 48, 49, *50*
prey taken, 87, 88, 89; bird species, 93, 94, 95, 96; fish species, 90; mammal species, 99, 100
protective legislation, 132
productivity of Sea Eagles, 137, 138
source of eaglets for reintroduction, 157, 158, 164–6
Norwegian Air Force,
    transport of eaglets, 194

Orkney,
    accessibility of eyries, 114
    bounty schemes for eagle destruction, 112
    nesting sites, 123
    records of Sea Eagles, 43–4; prehistoric, 108–9
Osprey (*Pandion haliaetus*), 90
    artificial nests and conservation, 134
    classification, 5, 6
    decline of, x, 128
    distribution of, 34, 38, 155
    eviction by Sea Eagles, 54
    in falconry, x
    fishing ability of, 79
    natural recolonisation by, 154, 155
    number required for population maintenance, 157
Otter (*Lutra lutra*),
    on Rum, 162
    and Sea Eagles: battles between, 36, 100; piracy of Otter's food, 90; prey of Sea Eagles, 100

Pallas' Sea Eagle (*Haliaeetus leucoryphus*), 8, *10*, 11
    distribution of, 13, 30; map, 26
    plumage patterns on underside, *12*
Pandion haliaetus, see Osprey
pellets,
    from captive Sea Eagles, 172
    composition of, and diet investigation, 84, 86, 165
    from newly released eagles, 159, 178
    regurgitation of, 84, *85*
    size of, 84, 85
Pennant, Thomas,
    on classification of Sea Eagle, 2
Peregrine (*Falco peregrinus*), 93
    harassment by Air Ministry, 128
    mercury in museum specimens of, 136
    in falconry, ix
    as pollution indicators, 136, 155, 202
    reintroduction in USA, 147–8
    Sea Eagle eyries used by, 54
pesticides, *see* toxic chemicals

Index

piracy by Sea Eagles, 88, 90
plumage,
 coloration and age, 18, *19, 20, 21*
 development of, 68
 and identification in the field, 3–4, *12,* 17–18, 186
 of immature, *189*
 moulting, 14–18, *19, 20, 21*; and individual bird identification, 18, *22*; of tails, 17; of wings, 15–17, *16*
 numbers of feathers, 14, 15
 of sea eagle species, 8, 9–13, *9, 10, 11*
 sexual dimorphism, 18
 weight of, 14
Poland, 26, 29–30
 brood size, 71, 72
 protection of eyries, 133
Portugal, 27
predation, *see* hunting

quarantine,
 and retention on Rum, 178
 and site of captivity, 164

radiotelemetry,
 and Sea Eagle tracking, 81, 184–6
recolonisation,
 breeding sites, 203–4, 206–7
 dispersal from Rum, 182, 192, 193, 194, 198, 203–4; map, 200
 number released, 180, 194, 197–9; and number required for population maintenance, 157–8
 and survival, 157, 197–9; to adulthood, 207, enhancement of, by previous releases, 178, 202
 (*see also* reintroduction)
Red Deer (*Cervus elaphus*),
 predation by Sea Eagle, 190–1
 on Rum, 162, 163; natural death of, and carrion provision, 162, 197
Red Kite (*Milvus milvus*),
 decline of, 128
 eviction by Sea Eagles, 54
 fondness for carrion, 128
 talon-grappling by, 55
reintroduction of bird species, 147–9, 151
reintroduction of Sea Eagles, 164–72
 attempts at: in Argyll, 158; in Fair Isle, 158–61
 cage site and design, 164
 collection of eaglets from Norway, 191, 192, 193, 164–6
 criteria for, 153
 definition of reintroduction, 153
 and Fulmars, 93, 117–18, 160
 by hacking, 156, 196
 loss of reintroduced birds, 180, 181, 183–4, 188, 193–4, 196–7, 199; due to illness, 166, 169, 170; due to poisoning, 193, 201–2
 measurements prior to release, 174, *175*; weights, 174, 180
 methods considered: fostering of eggs and chicks, 156; implantation of adults, 155; release from captive-bred stock, 157
 number required for population maintenance, 157–8; and number released, 180, 194, 197–9
 and public sympathy for, 153, 154
 radiotracking, 184–6
 and sheep farming, 101–5
 sites considered for, 161–2, 202
 source of Sea Eagle stock, 157, 158, 164–6
 tethering, 167, 169, *173,* 179; escape and recapture, 167, 169, 182–3
 versus natural recolonisation, 155
 (*see also* recolonisation; release; restocking)
release,
 from captive-bred pairs: in Germany, 143, 146
 dates of, on Rum, 171, 180, 181, 184, 185, 186, 188, 189, 191, 192, 193, 194
 flight on, 176, *177,* 178
 measurements at time of, 173, *175*; weights, 173, 180
 moment of, 159, 176, 177–8
 number of individuals, 180, 194, 197–9
 rate of, for population maintenance, 157
 and sex ratio, 180, 194
 with radio equipment, 185
restocking,
 with Bald Eagles in New York State, 148–9, *150*
 definition of, 152
 programmes for, 149
 (*see also* introduction; reintroduction of Sea Eagles)
Rhum, *see* Rum
ringing,
 British Trust for Ornithology ring, 174
 Darvic colour coded ring, 174
 and identification in the field, 186, 192, 193, 206
 prior to release, 174
 recoveries in Europe, 74–7
 (*see also* identification in the field)
Romania, 26, 28
 chick sibling aggression, 73
 diets of Sea Eagles, 86, 87
 nesting sites, 48
Royal Air Force, Kinloss,
 transport of eaglets from Norway, 164, 166, 193

Royal Society for the Protection of Birds (RSPB), xi, 166
  reintroduction, 163, 203; on Fair Isle, 158–61; on Rum, 163, 203
Rum, 162–3, *163*, 205
  decline of Sea Eagles, 41 115, 120, 121
  Golden Eagles on, 41, 163, 202; egg-laying period of, 59
  map of, 162
  restoration of habitats of Golden Eagle and Sea Eagle, 205
  spelling of, 2
  suitability of, for reintroduction, 161–2, 202

Sagittarius serpentarius, see Secretary Bird
Saker (*Falco cherrug*), 54
St Kilda,
  Fulmar on, 118
  Sea Eagle records, 43, 115
Sanford's Sea Eagle (*Haliaeetus sanfordi*), 8, 9
  distribution of, 13
  plumage, 8; patterns on underside, *12*
Sardinia,
  extinction of Sea Eagle, 27
Scotland,
  decline and destruction of Sea Eagles, 113, 114–15
  egg-laying period, 58–9, 60
  fledging, 74
  myths and legends, 3, 37, 109, *110*
  names for Sea Eagles, 1–2, (see also names)
  nesting sites, 48–9, 51–2, 126
  persecution, 118–20, 121, 122–3
  prey taken, 96, 100
  records, 33, 35–46, 112, 115; prehistoric, 109
  reintroduction attempt, 158
  sheep farming and Sea Eagles, 101–5
Scottish Wildlife Trust (SWT), xi, 166
Secretary Bird (*Sagittarius serpentarius*), 5, 6
sexual dimorphism,
  in Sea Eagle: calls, 18; plumage, 18; size, 18–19, 22–4, 174; weight, 18–19, 23–4, 173
sheep farming, 118–19
  improved husbandry *versus* predator control, 104–5, 119
  and Sea Eagle reintroduction, 101–5; and attitudes of farmers, 26–7, 101, 125; and safety of lambs, 102, 120, 160, 201
Shetland,
  decline and destruction of Sea Eagles, 115, 120; bounty schemes for, 112
  fish species taken, 91
  lambs and Sea Eagles, 102
  nesting sites, 49, 123

records, 13, 44–5, 108, (of albino) 13, 45, *46*
Shiant Islands,
  desertion of eyries, 115
  long tenancy of eyrie, 42, 51
sight, sense of, 81
size of Sea Eagle, 4
  and north to south distribution, 24, 63, 64
  sexual dimorphism for, 18–19, 22–4, 173
  wing span, 81
skulls, of predatory birds, *80*
Skye,
  clutch size, 61
  combat between Sea Eagles, 55
  decline and destruction, 41–2, 115, 121
  fish species taken, 91
smell, sense of, 81
snake eagles,
  classification of, 5, 6
Spain, 27
Spanish Imperial Eagle (*Aquila heliaca*), 54, 147
Sparrowhawk (*Accipiter nisus*), 93
  decline of, 128
  sexual dimorphism for size, 19
  talon-grappling by, 55
  and toxic chemicals in food chain, 136
Steller's Sea Eagle (*Haliaeetus pelagicus*), 8, 10
  distribution of, 13, 30; map, 26
survival,
  enhancement of, by previous releases, 178, 202
  of reintroduced Sea Eagles, 157, 197–9; to adulthood, 207
Sweden, 26
  brood sizes, 72
  chick sibling aggression, 72–3
  conservation, 134; artificial eyrie construction, 134; cross-fostering chicks, 146; forestry practices, 133; Nature reserves, 133; protective legislation, 31, 132; uncontaminated food provision, 139, *140*
  decline of Sea Eagles, 31; and pesticides, 136–8
  destruction of Sea Eagles, 108
  dispersion from, 74, 75, 76
  egg-laying period, 60
  nesting sites, 48
  prey taken, 87, 100
  productivity, 137, 138
SWT, *see* Scottish Wildlife Trust
symbolic eagles, 2, 108–9
  (*see also* myths and legends)
Syria, 26, 28

talon-grappling, *see under* flight